BK 633.1 M446C
CEREAL SCIENCE
/MATZ, SAMU
1969 15.00 FV

3000 298218 30015
St. Louis Community College

WITHDRAWN

633.1 M446c   FV
MATZ
  CEREAL SCIENCE
                              15.00

JUNIOR COLLEGE DISTRICT
of St. Louis - St. Louis County
LIBRARY
5801 Wilson Ave.
St. Louis, Missouri 63110

 PRINTED IN U.S.A.

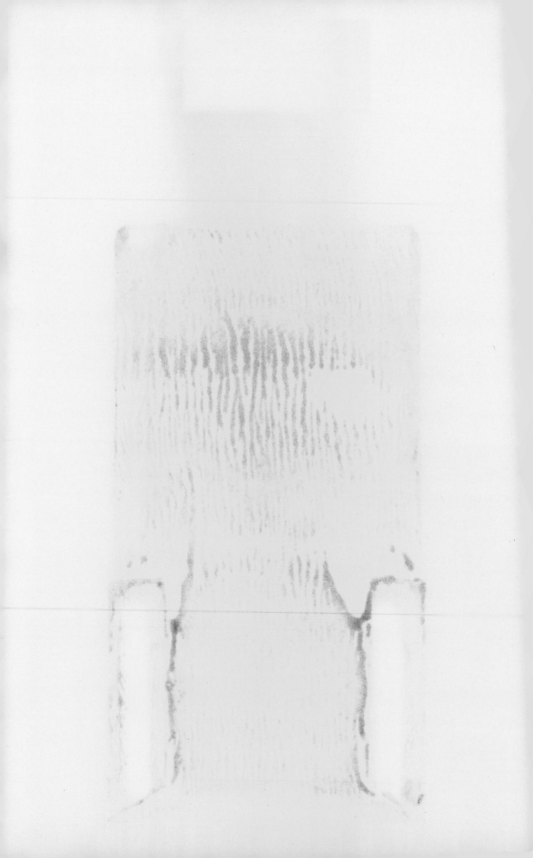

other AVI books

Cereal Science

by SAMUEL A. MATZ, PH.D.

Vice President

*Robert A. Johnston Company, Milwaukee, Wisconsin.
Formerly, Technical Director Refrigerated Dough Program,
The Borden Foods Co. At one time Chief, Cereal and
General Products Branch, Quatermaster Food and Container
Institute for the Armed Forces; Chief Chemist, Harvest
Queen Mill and Elevator Company, Plainview, Texas;
Instructor, Department of Flour and Feed Milling Industry,
Kansas State University*

WESTPORT CONNECTICUT

THE AVI PUBLISHING COMPANY, INC.

1969

Copyright 1969 by

THE AVI PUBLISHING CO., INC.
Westport, Connecticut

REGISTERED AT STATIONERS' HALL
LONDON, ENGLAND 1969

All Rights Reserved

Library of Congress Catalog Card No. 79-75557

SBN 87055–061–6

Printed in the United States of America

This book is based on the first eight chapters of *The Chemistry and Technology of Cereals as Food and Feed* which was published in 1959. All of the chapters were rewritten to bring them up-to-date and, in most cases, they were expanded to include additional material on the composition and the nutritional qualities of the grains. The remaining parts of the 1959 volume will be included in revised form in *Cereal Technology* to be issued later.

Only cereal grains are discussed in the present book, and so buckwheat (an herbaceous plant), soybeans (a legume), and some other nongrasses bearing seeds which are commonly processed and consumed similarly to cereal grains, are not included. Furthermore, the forage uses of cereal plants are touched upon only briefly. The material on production, trade statistics, and culture was thought to fit naturally into the discussions of the grains even though including it as part of a book entitled *Cereal Science* might be rather unexpected to some readers.

So far as culture is concerned, the greatest emphasis has been placed on practices followed in the United States although important variations of these procedures followed in other countries have also been discussed.

Most of the contributors to the earlier volume agreed to submit revised chapters, but in some cases the final version was a collaborative effort. Dr. Stanton, who wrote the 1959 Chapter on Oats, passed away in the interim, and I prepared a new manuscript having a considerably different emphasis but incorporating part of the original material. H. M. Beachell, who retired from government service in 1963, is engaged in full-time research at the International Rice Research Institute in the Philippines and did not have the time or facilities to prepare a revision of his Chapter on Rice. I rewrote his chapter, incorporating more information and bringing it up-to-date.

Dr. Kramer was also unable to supply a revised manuscript on sorghum due to the pressure of his other duties, so I prepared a new version which he reviewed.

The final Chapter, containing brief discussions on millet, wild rice, adlay, and rice grass, remains much the same as it was in the earlier publication.

Many people assisted our work by supplying data, suggestions, illustrations, etc. I especially wish to thank Dr. E. F. Caldwell and the Quaker Oats Co. for furnishing photomicrographs of various grains as well as for permitting access to unpublished information on oats. Dr. Caldwell also reviewed the Chapter on Oats. Much advice on the Corn Chapter and some illustrative material was supplied by Dr. S. A. Watson of the Corn Products Co.

Contents

CHAPTER 1

John A. Shellenberger | **Wheat**

HISTORICAL ASPECTS OF WHEAT PRODUCTION

Origin of the Wheat Plant

Despite many years of intensive effort and extensive investigation, it has not been determined accurately either when or where the first cultivated wheat originated. At the beginning of recorded history, wheat was already an established crop whose origin was unknown (Anon. 1953).

The ancestry of the common races of wheat grown today remains problematical; however, a good deal of evidence indicates that cultivated einkorn was developed from a type of wild grass native to the arid pasture lands of southeastern Europe and Asia Minor.

Emmer, which is generally regarded as one of the ancestors of today's wheat, closely resembles a wild species of wheat found in the mountainous regions of Syria and Palestine. It is a much better wheat than einkorn, which gives low yields and dark, somewhat bitter kernels. Since crude wheat-type plants, like einkorn and emmer, and many wild species of grass were growing centuries ago, Percival (1921) concluded that bread wheat originated by hybridization from an emmer type and a wild species of grass.

Advances in nuclear physics have helped immensely to fix the dates in history for the beginning of agricultural plants. While plants are alive, they absorb from the atmosphere carbon dioxide which contains traces of radioactive carbon 14. When plants die, the supply of radioactive carbon ceases, because carbon dioxide is no longer being ingested. Since carbon 14 slowly disintegrates, and since the rate of loss of radioactivity is known, the time of death can be established by determining the ratio between the radioactivity of the carbon in living and in dead plants.

Evidence of this kind places the beginning of the cultivation of wheat roughly 6,000 yr ago. Cultivation began in the Syria-Palestine area and spread west and south into Egypt and east into Iran. From Iran, wheat spread into India, China, Russia, and Turkistan; from Egypt, Palestine, and Syria, it moved into South

JOHN A. SHELLENBERGER, is Distinguished Professor, Department of Grain Science and Industry, Kansas State University, Manhattan, Kan.

1

Courtesy of Agronomy Dept., Kansas State Univ.

FIG. 1. THE LESS COMMON SPECIES OF WHEAT COMPARED WITH *Triticum vulgare*, THE SPECIES COMMONLY GROWN

Left to right, spikes of *T. vulgare*, club, einkorn, emmer, durum, spelt, poulard, and Polish.

and Central Europe.   The first wheat to reach Europe was in the form of einkorn and emmer about 3000 BC.   Bread wheats, such as we know today, began to spread over Europe from southern Russia about 2000 BC.   The less common species of wheat compared with the types grown today are shown in Fig. 1.

Cultivation of Wheat

For centuries, agriculture was so primitive that any attempts to describe its beginnings become a matter of surmise and conjecture. The history of the start of cultivation of cereal crops far precedes any documents;   therefore, it can only be supposed that the use of certain grains for food led primitive man to develop the fundamental art of scattering seeds on rough clearances of the soil.   Gradually, crude methods for cleaning land and cultivating the soil developed. The earliest discoveries of archeologists show that even neolithic man had wheat grains not much different from the types known today. With the advent of written history at the time of the Greeks and Romans, agriculture was a highly developed art (Storck and Teague 1952).

Once the avenues of commerce were formed, wheat culture spread rather rapidly throughout the world.   Wheat was unknown in both North and South America until brought to the Continent by early explorers from Europe.   Because the wheat plant is extremely adaptable to environmental conditions, wheat production encircles the globe.   Harvesting is being done somewhere on the earth every

TABLE 1

THE WORLD'S WHEAT HARVEST

Month	Country	Month	Country
January	Australia	July (Contd.)	USSR (South)
	New Zealand		Switzerland
February	Egypt (Upper)		United States
March	India	August	Belgium
April	Middle East		Canada
	Egypt (Lower)		Denmark
	Mexico		England
May	China		USSR (Central)
	Japan		Poland
	Morocco		Netherlands
	United States (Texas)		United States (North)
June	France	September and	Norway
	Greece	October	USSR (North)
	Italy		Sweden
	Spain	November	Africa (South)
	Turkey		Peru
	United States	December	Australia
July	Austria		Argentina
	Hungary		Africa
	Bulgaria		Chile
	England		New Zealand

month of the year, as shown in Table 1. Wheat is grown from the latitude close to the Arctic Circle in the Northern Hemisphere, both in Canada and in the Scandinavian countries, and beyond the 40th parallel in South American and New Zealand.

The cultivation of wheat varies from primitive methods, still practiced in a few areas of the world, to complete mechanization of the entire operation from soil preparation to harvesting. In the more important wheat-producing areas land is plowed, prepared for seeding, and sowed using tractor equipment, after which the crop is harvested by self-propelled combines and the wheat transported to storage or to railway terminals by truck.

BOTANY OF THE WHEAT PLANT

Structure of the Wheat Kernel

Percival (1921) has described the wheat kernel in great detail. The kernel or seed of the wheat plant is a nut-like fruit called by botanists, a caryopsis. It contains a single seed or kernel enclosed within a thin shell, and the seed cannot, as is true with some fruits, be separated readily from the shell, or pericarp.

The main features of the wheat kernel can best be described in terms of the rounded or dorsal side and the ventral or crease side. A deep groove or crease extends the entire length of the wheat kernel.

At the apex or small end of the grain there are many short fine hairs known as brush hairs. The outer bran or seed coat consists of three layers known as the epidermis, epicarp, and endocarp. The other portions of the kernel are the germ and endosperm. Wheat kernels vary considerably in size, form, and color.

From a processing standpoint, milling separates the three main parts of the wheat kernel into products commonly known as flour, bran, and germ. The wheat type, chemical composition, and physical structure, combined with the equipment used and skills applied, determine the success of the milling operation.

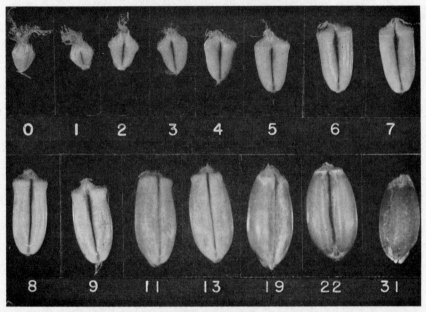

Courtesy of G. E. Scott, E. G. Heyne, and K. F. Finney

FIG. 2.   RATE OF DEVELOPMENT OF HARD RED WINTER WHEAT KERNEL

Numerals indicate number of days after pollination.

Detailed studies of the structure of the mature wheat kernel have been reported by Bradbury *et al.* (1956, 1956A, and 1956B). External, transected, and longitudinally bisected views of the wheat kernel as described by those authors are shown in Figs. 4, 5, and 6. They report the bran to be composed of the pericarp and the outermost tissues of the seed, including the aleurone layer, and also that there is no line of cleavage between the bran and starchy endosperm.

That accounts for some of the difficulties encountered in flour milling when a complete separation of the two is desired. The germ, however, is a separate entity, and no breakage of cell walls is required to separate germ from endosperm.

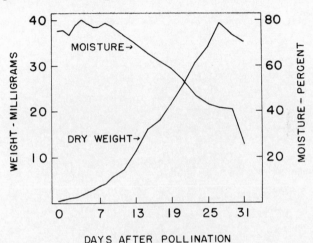

DAYS AFTER POLLINATION

Courtesy of G. E. Scott, E. G. Heyne, and K. F. Finney

FIG. 3. DRY WEIGHT AND MOISTURE CONTENT OF HARD RED WINTER WHEAT KERNELS HARVESTED AT VARIOUS DATES AFTER POLLINATION

The various parts of the wheat kernel are named and shown in the illustrations. The tissues of the pericarp form a thin protective covering over the entire wheat kernel. Microscopic examination of the pericarp shows that it is composed of several layers, named from the outside of the kernel inward as follows: epidermis, hypodermis, remnants of thin-walled cells, intermediate cells, cross cells, and tube cells. The hairs of the brush are extensions of epidermal cells. Any damage to the protective covering of the kernel, particularly if it permits moisture loss or gain, is important to storage considerations. Bradbury *et al.* (1956) have presented data to show that from 25 to 73% of the kernels of commercial lots of wheat had endosperm or germ exposed as a result of mechanical injury to the pericarp or seed coat, which explains difficulties in cleaning, tempering, and milling wheat.

Germination and Growth

Development of the wheat plant starts with the germination of the seed. Before sown seed can germinate and growth proceed,

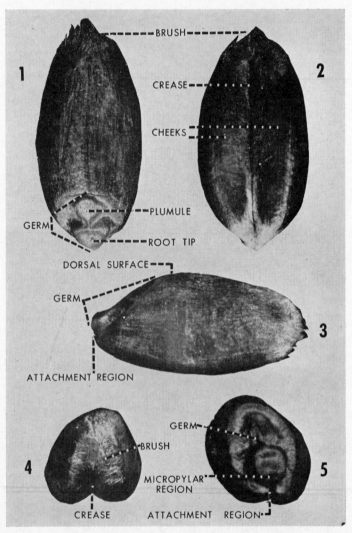

Courtesy of US Dept. Agr.

FIG. 4. EXTERNAL VIEWS OF A TYPICAL HARD RED WINTER
WHEAT KERNEL

1—Back (dorsal) face; 2—crease (ventral) face; 3—side; 4—brush;
5—germ end.

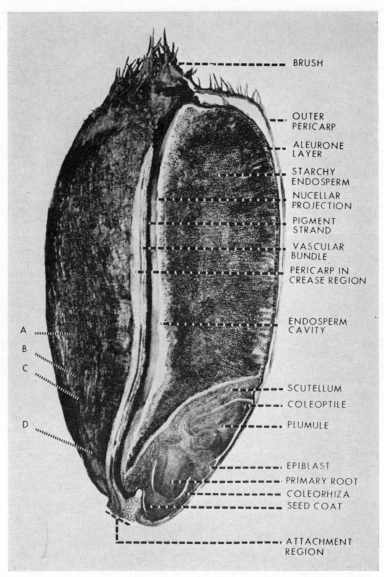

BRUSH

OUTER
PERICARP

ALEURONE
LAYER

STARCHY
ENDOSPERM

NUCELLAR
PROJECTION

PIGMENT
STRAND

VASCULAR
BUNDLE

PERICARP IN
CREASE REGION

ENDOSPERM
CAVITY

A

B

C

SCUTELLUM

COLEOPTILE

D

PLUMULE

EPIBLAST

PRIMARY ROOT

COLEORHIZA

SEED COAT

ATTACHMENT
REGION

Courtesy of US Dept. Agr.

FIG. 5.   VIEW OF WHEAT KERNEL BISECTED LONGITUDINALLY

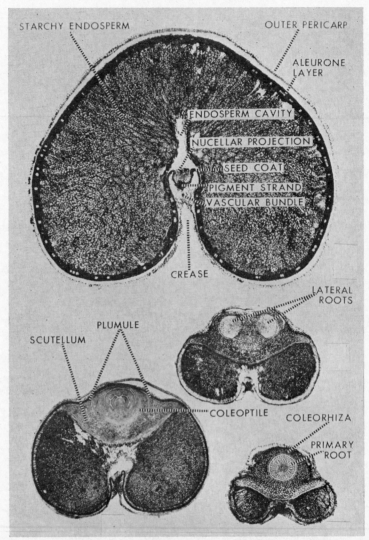

Courtesy of US Dept. Agr.

FIG. 6.   TRANSECTION OF WHEAT KERNEL

several conditions must be fulfilled, such as: (1) an adequate amount of moisture in the soil, (2) sufficient soil warmth, and (3) a supply of oxygen.   Lack of water, heat, or air will prevent germination.

The wheat kernel, after being sown, absorbs moisture and becomes swollen.   Soon the pericarp at the germ end of the kernel ruptures and the plumule of the embryo emerges. Later the primary root emerges, followed a few hours later by the first pair of lateral rootlets. The roots grow and expand to provide the plant with water and soil nutrients. The acrospire develops and pushes up through the soil to emerge as a shoot, forming the first foliage. The leaves of grasses (wheat is classified as a grass) are attached to the joints of the stem alternately, in a manner that produces two opposite longitudinal series of leaves. At the base of the leaf-blade of wheat, a membrane structure nearly surrounds the stem. Leaves grow in length from the base area near the stem; therefore, grazing or cutting back the leaves does not prevent renewed growth.

As the wheat plant develops, the stem extends, and lateral buds formed near the surface of the soil grow into short stems, thus producing a branching close to the ground called "tillering." The "tillers" later form the straws of the plant and can vary in number up to 100 from a single kernel. The inflorescence or ear of wheat consists of the rachis with spikelets on alternate sides. At the base of the spikelets are two chaffy scales called glumes, followed by a number of flowers arranged along the rachilla, and terminated by the beard or awn.

When the ear or head emerges from the upper leaf sheath, flowering takes place. The first spikelets to flower are usually toward the middle of the ear, and the whole ear completes flowering under normal weather conditions in 5 or 6 days. After fertilization, the grain begins to develop in volume, increasing as illustrated in Fig. 2.

Work on the daily development of the wheat kernel was reported by Scott (1955), Scott et al. (1957), and Jennings and Morton (1963A). The dry weight and moisture content of wheat kernels harvested at various dates after pollination were studied, as shown in Fig. 3. The dry weight of individual kernels increased slowly for the first few days after pollination, but thereafter increased rapidly until the kernel appeared physiologically mature on the 27th day. Kernel moisture content reached a maximum on the 15th day, then decreased gradually until the 29th day, when rapid loss of water began.

Chemical Composition During Kernel Growth

The development of the wheat kernel has been studied extensively and agronomic research is reported in the book edited by Quisenberry and Reitz (1967), but interesting investigations concerning changes in chemical composition have also been conducted. Jennings and Morton (1963A, 1963B, and 1963C) have followed changes in wheat kernel carbohydrates, nitrogen compounds, and nucleic acids from time of flowering. When reported on a dry matter basis, sucrose, reducing sugars, and pentosans all decrease rapidly in the endosperm with time from flowering, while the starch content rapidly increases. On the same dry matter basis, total nitrogen decreases rapidly at first but then increases slightly; protein nitrogen decreases slightly for the first few days and then slightly increases; the nonprotein nitrogen decreases during the entire period, and the amide nitrogen remains nearly constant.

Changes in nucleic acids and other phosphorus-containing compounds while the wheat kernel develops have been studied by the same authors. Changes in DNA content of the endosperm indicate rapid cell division during the first 14 days after flowering. After that kernel development was due to cell expansion, rather than new cell formation. It must be recognized that the fruiting period for different environments varies greatly; and therefore, the time after flowering is not indicative of a fixed physiological state.

Much attention is being given to the biosynthesis of proteins in the developing wheat kernel and to the fractionation of protein by electrophoresis and column chromatography. Also, changes in protein with wheat kernel development have been followed by radioactive ^{15}N, ^{14}C, and ^{35}S-labeled techniques. Pomeranz et al. (1966) and Hoseney and Finney (1967) have reported on the amino acid composition of maturing wheat. The number and concentration of free amino acids decrease during wheat maturation. Several nonprotein amino acids were found at early stages of maturity. Finney (1954) and co-workers have for years pioneered in studies of protein synthesis and bread baking properties as the wheat kernel matures. Gluten formation adequate for some loaf volume potential commences as early as 21 to 24 days preripe and reaches an optimum 10 to 14 days preripe.

The lipids in maturing wheat kernels also have been studied (Pomeranz 1967). Wheat development is accompanied by a slight decrease in lipid content on an as is basis, but lipid content nearly doubles on a kernel basis, according to Daftary and Pomeranz (1965). Free fatty acids in mature wheat are less than half those in wheat 21–23 days preceding ripeness. No consistent changes are found in the quantity of polar lipids, and there is no correlation between lipid content and breadmaking potentialities.

The lipids have an important influence on baking qualities of flours, even though present in relatively small amounts. Phosphorus-containing lipids must be present for gluten to exhibit its normal rheology properties. However, as discussed by Pomeranz (1967), the lipids easily removed from flour by nonpolar solvents, such as ether or petroleum ether, affect breadmaking or rheological properties very little. Small amounts of polar lipids improved bread quality of flours, while the nonpolar components had little effect. Adding nonpolar components to a defatted flour has a deleterious effect on bread quality. At the present stage of our knowledge, the role of wheat lipids in flour use is not fully understood.

Finney (1954) has shown that high temperatures and low humidities during fruiting of wheat the last 15 days before ripeness can result in subnormal physical dough properties and reduced loaf volume. The relationship between high temperatures (above 90°F) during maturation and impaired baking properties is complex with many variables involved.

CLASSIFICATION OF WHEAT

Species

Wheat belongs to the grass family, Gramineae (Poaceae), and the genus *Triticum*, and until recently classification was almost entirely at the species level and below. When it was discovered that 2 of the 3 genomes of hexaploid wheat were derived from the genus *Aegilops*, considerations of revised classifications became necessary because an intergeneric hybrid cannot be included in the genus of one of its parents. Several different classifications have been suggested, but until this subject has been resolved the commonest and traditional grouping based on botanical characteristics can be used. Percival (1921) described 18 species, but only a few are agriculturally important.

The important species of *Triticum* follow:

T. aegilopoides and *T. monococcum* possess perhaps seven pairs of chromosomes and are somewhat more grass-like in appearance than are the common wheat varieties. The kernels are small, flinty, and rice-like.

T. compactum, or club wheat, has 21 chromosomes. This species is characterized by club-shaped spikes. The kernels are plump, soft, and mealy and may be either white or red.

T. dicoccoides has 14 chromosomes and is a form of wild emmer. This species has certain disease-resistance qualities and has been used successfully in crosses with hard red spring wheat varieties. The kernels are long, narrow, and flinty.

T. durum has 14 chromosomes. It is often referred to as macaroni wheat, because hardness of the endosperm suits it for manufacturing of macaroni and related products.

T. polonicum, known as Polish wheat, has 14 chromosomes. The grains are flinty, long, and narrow. This species is not of great commercial importance.

T. spelta or speltz has 21 chromosomes and retains its glumes as does emmer with which it is often confused. The kernels are long and flinty. This species has no economic importance in the United States.

T. timopheevi has 14 chromosomes It is a fairly recent species possessing considerable disease-resistant qualities.

T. turgidum or cone wheat has 14 chromosomes. The kernels are large, plump, and usually mealy.

T. vulgare is a 21-chromosome wheat commonly referred to as a bread wheat. It varies in size of kernel, can be either flinty or mealy, and can have either spring or winter habit. Most of the wheat varieties used for bread-baking are of this species.

Hybrid Wheat

Modern wheat improvements have been sought principally through the process of crossing two or more wheat varieties and selecting from the progeny the desirable characteristics. Although such wheat breeding has commonly been referred to as producing hybrids, it does not qualify as a procedure for evolving true hybrids. Because of the adaptability to environment and yield potentials exhibited by hybrid corn and sorghum, there is intense research activity by agricultural experiment stations and industrial scientists to produce a true wheat hybrid that involves cross pollination in the field.

The development of a hybrid wheat depends on: (1) a male sterile parent, i.e., one that does not shed pollen; (2) a male parent capable of restoring fertility to the male flower parts in the field; and (3) pollination of the male sterile wheat by a fertility-restoring male parent. Through the work of Fukasawa (1958), Kihara (1961), Wilson and Ross (1961) and others, all three of the aforementioned steps have been accomplished, and full hybrid wheats have been grown. The task of incorporating desirable characteristics into the sterile and restorer cytoplasms and quality testing programs will require time before all tools available are used effectively.

The development of hybrid wheat will have a profound effect on the entire wheat economy, and a bright future seems ahead where there are possibilities for increased production and wheat characteristics changes made to best suit the needs for processing and nutrition.

Methods for Distinguishing Wheat Varieties

It is difficult or impossible to identify, within a class, most varieties of wheat by examining the seed. Field appearance and all plant characteristics usually need to be studied before identification is possible. When need arises, it has proved feasible to present in a simplified, nontechnical manner, based on morphological characteristics, methods to identify some varieties of wheat. An example is identification of hard red winter wheat varieties as an aid to wheat growers, grain dealers, and millers. Several publications on the subject of kernel identification have appeared, one of the most recent being Extension Circular No. 254, Kansas State University (Anon. 1956).

Identification is based on kernel characteristics such as: color,

Courtesy of US Dept. Agr.

FIG. 7. DISTINGUISHING WHEAT HEAD AND KERNEL CHARACTERISTICS OF
PONCA (LEFT) AND RED CHIEF (RIGHT)

These are examples of good and poor quality varieties.

texture, shape, germ, back, crease, and brush, plus other considerations of wrinkling, depressions, fine lines, or sharply outlined germ. Certain varieties less desirable from the standpoint of baking quality include Red Chief and Chiefkan which can be distinguished from Turkey-type wheats easily by field appearance, since the former varieties are beardless. Kernel characteristics are also sufficiently different to be recognizable by persons who have had some instruction in wheat kernel identification. Figure 7 illustrates the distinguishing characteristics of Red Chief kernels compared with Ponca, one of the wheat varieties grown in the Southwest.

The need for variety identification has been intensified by the US Dept. of Agr. establishment of a list of undesirable wheat varieties subject to discount under its loan program to farmers. Several undesirable varieties are listed for the following classes: Hard Red Winter, Hard Red Spring, White, Soft Red Winter, and Durum.

PRODUCTION STATISTICS

The Uniqueness of Wheat

Wheat has been for centuries, along with rice, the staple cereal food for man. The principal rice consuming area is Asia, where rice's use as a food involves relatively simple processing and cooking procedures. In contrast, the processing and multiplicity of products produced from wheat have created major demands for specific quality characteristics and nutritional values. Wheat composition and quality are the results of the interaction of (1) environment, (2) soil, and (3) variety.

The uniqueness of wheat among cereal grains lies in the character of its protein content. As a plant develops from a seed, two metabolic processes are taking place in the cells, namely, photosynthesis and nitrogen fixation. Photosynthesis involves formation of carbohydrates from carbon dioxide, water, and energy while nitrogen fixation is the conversion of gaseous nitrogen into combined nitrogen that can be readily assimilated by the plant. Nitrogen fixation as discussed by Stewart (1967) can be carried out by root nodule-bearing legumes, certain algae, by chemical fertilizer industries, or by an electrical discharge in the atmosphere.

Numerous studies have been conducted comparing nitrogen sources on yield and quality of wheat. It is difficult to predict the amount of available nitrogen a given soil will contribute to a wheat crop; however, it is recognized that nitrogen fertilizer has led to greatly improved grain yields and to increasing protein content of the wheat kernel. The task that lies ahead is for the scientist to learn more about the changes occurring in the developing plant, and to control the processes for greater efficiencies in putting together the desired amount and quality of protein in the wheat kernel.

The biochemistry of the proteins of wheat has been reviewed from time to time as new knowledge or new interpretations of existing facts develop. A recent review of wheat protein research during the past 50 yr prepared by Sullivan (1965), and Wall (1967) has given a good review of the origin and behavior of flour proteins. The proteins of wheat are complex, and there is no simple explanation of their constitution or biological function. Neither differences in the amounts of the various classes of proteins nor differences in the amount or kind of amino acids account for the wide variations in rheological and baking properties of flours. From a practical viewpoint, the great weakness in the multitude of excellent approaches to the characterization of wheat proteins has been lack of proof of direct association with quality.

Recent research by Hoseney *et al.* (1968B) has related protein solubility and electrophoresis studies to bread baking properties. It was shown conclusively that the water-soluble fraction of flour is not responsible for quality differences; however, the water-soluble fraction is required to produce a normal loaf of bread. The water solubles were found to have a dual role by (1) contributing to gassing power, and (2) modifying the physical properties of the gluten. The dialyzable fraction of the water solubles contributed to gas production and the glycoproteins are involved in the modification of the gluten.

The factor or factors responsible for baking quality differences reside in the gluten protein. Ultracentrifugation of gluten solutions ($0.005N$ lactic acid) at $100,000 \times G$ separated the proteins too large to enter the starch-gel (glutenins) from the proteins migrating into the starch-gel (all the gliadin plus 65% of the glutenins). Those two fractions, when reconstituted and baked into bread, gave loaf volume and crumb grain equal to that of the original flour. Interchanging the two fractions from good and poor quality varieties showed that the proteins migrating into the starch-gel were responsible for loaf volume potential. Further fractionation of the proteins migrating into starch-gels by either ultracentrifugation at higher R.C.F. ($198,000 \times G$) or precipitation with 70% ethyl alcohol yielded fractions that replaced the nonmigrating proteins in baking.

Starch-gel electrophoresis patterns of the above fractions showed that a significant part of the glutenins migrated into the starch-gel in the absence of any gliadin proteins.   As much as $^2/_3$ of the glutenin proteins migrated into the starch-gel in the presence of the gliadin proteins, which is evidence of a glutenin-gliadin interaction in gluten. As additional evidence, starch-gel electrophoresis and gel-filtration showed that high molecular weight proteins increased during gluten formation.

Starch, the other major constituent of the wheat kernel, is formed by the process of photosynthesis and is deposited in plant cells as granular particles.   Many genes are involved in determining the shape, crystalline pattern, and chemical properties of starch granules. Developments in the carbohydrate field are too numerous to be dealt with here, but the starch-protein matrix and the size of the starch granules in the wheat kernel are important in flour milling and baking processes.   Sandstedt (1965) prepared a review of 50 yr of progress in starch chemistry that covers the subject well, especially the cereal starches.

Farrell *et al.* (1967) have provided extensive analyses of US grown wheats and the products milled from wheat.   Another paper, Waggle *et al.* (1967), gives quantitative amounts of amino acids, vitamins, minerals, and gross energy of the same samples.   Thus the chemical composition of the wheat kernel for different classes is fully established.

World Wheat Production

It is difficult to obtain a completely reliable picture of the quantity of wheat grown in various parts of the world because of continually changing conditions.   Somewhere in the world, wheat is being planted or harvested throughout the year;   consequently, there is no ideal time to choose as a basis for estimating total production.   Another complication is absence of data on wheat production from Mainland China.

Wheat is grown to some extent on all six continents, but as shown in Table 2, about $^9/_{10}$ of the total is grown in Europe, Asia, and North America (Quisenberry and Reitz 1967).   Contrary to popular belief, even though Asia is considered a rice producing and consuming area, wheat production is high there, averaging in recent years about 1.8 billion bushels.   World production of wheat, including estimates for China and USSR, has increased to over 8 billion bushels in the past half century.   Present wheat production in Asia exceeds that in either North America or Europe, excluding the Soviet Union. The three leading wheat producing countries are the United States, the Soviet Union, and China.   The principal wheat producing coun-

TABLE 2

WHEAT PRODUCTION BY CONTINENTS[1]

	Average Production			
Continent	1960–64 (×1,000 M.T.)	1965 (×1,000 M.T.)	1966 (×1,000 M.T.)	1966 (1,000 Bu)
Europe	56,284	67,286	62,382	2,293,000
North America	49,505	55,496	60,233	2,213,000
Asia	51,119	55,715	51,096	1,877,400
Oceania	8,546	7,350	12,510	459,700
South America	9,467	8,576	8,893	326,800
Africa	5,689	6,067	4,772	175,300
USSR (Europe and Asia)	50,000	46,500	75,000	2,755,000

[1] From US Dept. of Agr., Agr. Statistics 1967.

TABLE 3

PRINCIPAL WHEAT-PRODUCING COUNTRIES OF THE WORLD

Country	Production in Bushels (×1,000)	Country	Production in Bushels (×1,000)
North America		Oceania	
Canada	844.4	Australia	448.5
United States	1,310.6	New Zealand	11.2
Europe		Asia	
Bulgaria	116.6	Afghanistan	74.3
France	414.2	China	738.5
West Germany	166.4	India	393.9
Greece	72.1	Iran	117.2
Italy	345.6	Pakistan	145.9
Poland	127.5	Turkey	301.3
Romania	191.1	Africa	
Spain	176.9	Algeria	26.5
USSR	2,775.0	Morocco	29.8
United Kingdom	130.5	South Africa	20.5
Yugoslavia	169.8	United Arab Republic	59.5

tries are listed in Table 3. World wheat production areas are shown in Fig. 8.

Production in the United States

Wheat is grown widely throughout the United States as shown in Fig. 9. The principal wheat types by areas are hard spring wheat in the Northwestern states, hard winter wheat in the Southwest, hard and soft wheat in the Pacific Northwest, soft wheat in the area east of the Mississippi River, and durum wheat in North and South Dakota.

The most rapid expansion of wheat acreage harvested in the United States came early in the present century (Pool 1948). The average harvested acreage between 1899 and 1901 was nearly 51 million and it remained near that until 1913 when, due to World

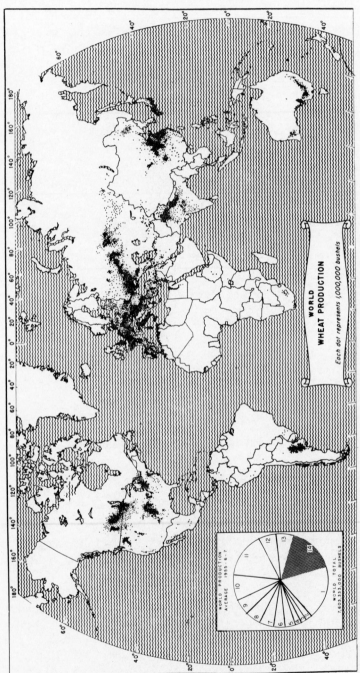

Fig. 8. World Wheat Production

Circle segments show share of world wheat production by countries: 1. USSR 24%;   2. Pakistan 1.6%;   3. West Germany 1.7%;   4. Australia 2.1%;   5. Spain 2.1%;   6. Argentina 2.8%;   7. Turkey 2.8%;   8. India 4.1%;   9. Italy 4.1%;   10. Franc 4.5%;   11. Canada 5.8%;   12. China 11%;   13. United States 13.7%;   14. All others, 19.7%.

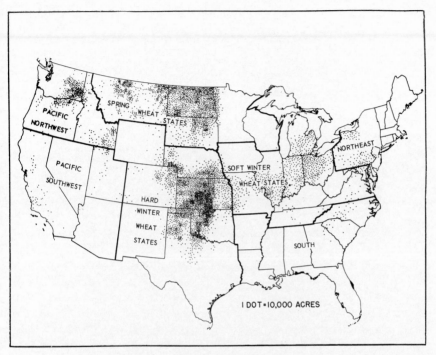

Courtesy of US Dept. Agr.

FIG. 9. WHEAT ACREAGE MAP OF THE UNITED STATES

War I, a rapid expansion culminated in planting 73 million acres in 1919.   After World War I, the harvested acres of wheat returned to a level near where they were at the start of the century, but acreage increased again during and after World War II.   The highest harvested acreage was in 1949, nearly 76 million acres.   Acreage allotments, the Soil Bank, and various governmental efforts to restrict wheat production have in past years reduced harvested acreage to well below 50 million acres, but reduced acres do not mean reduced wheat production.   Wheat acreage allotment in 1967 was increased to 68.2 million acres, largest since 1953, with no acreage diversion provisions.   On several occasions in past years, total US wheat production has exceeded one billion bushels, and in each recent year wheat has exceeded that value.   Several years Kansas has harvested over 200 million bushels, and North Dakota, 150 million bushels, together totaling more than $1/3$ of the entire production in the United States. South Dakota, Montana, Washington, Texas, Oklahoma, and Nebraska are other large wheat producing states.

TABLE 4

UNITED STATES WHEAT SUPPLY AND DISTRIBUTION, IN BUSHELS FOR YEAR 1965[1]

Total production	1,326,747,000
Used for seed	64,238,000
Milled for flour	511,000,000
Grain exports	639,260,000
Fed to livestock	64,060,000
Total disappearance	1,376,142,000

[1] Agricultural Statistics 1966, US Dept. of Agr.

The average US wheat supply and distribution during 1965 are shown in Table 4.

PRODUCTION METHODS

Growing the Crop

Farm practices for preparing the seed bed, seeding, and harvesting wheat differ from one section of the country to another. The most critical conditions prevail, usually in the semiarid Southwest, where land preparation evolves around moisture conservation. The general practices of preparing land for seeding to wheat follow the usual sequence of plowing, disking, harrowing, and drilling. Those operations are performed in various ways by common types of equipment. The end result sought is to prepare soil for the highest possible germination of seed and the production of healthy plants. Irrigation of wheat in the United States is increasing yearly. In Texas, nearly $1/3$ of the crop is grown under irrigation, and the acreage of irrigated wheat in Kansas and Oklahoma is large. Yields of over 200 bu per acre have been produced in Washington under irrigation.

Hard Red Winter Wheats

It is a general rule throughout the world that wherever winter wheat can be grown, it will be. Planting in the fall, and thus permitting the plant to establish a root system before the dormancy period, makes it possible for the plant to effectively use spring moisture, warmth, and sunshine. There is no spring delay caused by waiting until the field becomes sufficiently dry to cultivate. Thus, except in areas where the loss from winter killing is great, it is common to produce winter wheat. In most instances, there are definite quality differences between wheats of winter and spring habit, but that is not the universal rule. There are areas where seasons permit either spring or winter wheat to be grown. For example, in the Pacific Northwest it is customary to sow winter wheat, but if the stand is poor because of winter killing or poor germination due to dryness, it is possible to reseed in the spring, and with favorable weather conditions, produce a spring wheat crop.

Varieties of hard red winter wheat are grown under widely varying environmental conditions ranging from the eastern to the extreme western parts of the United States, and in all but the coldest regions. However, even in areas of extreme cold winter weather, such as Montana, considerable winter wheat is grown.   The southwest states are the principal producers of hard red winter wheat, with Kansas the leading state.   Shellenberger (1957) reviewed the history of wheat production in Kansas, the state that produces the most wheat.

In many sections of the Southwest where moisture is marginal, it is customary to alternate seeding with a year of summer fallowing to conserve moisture.   Much of the wheat farming in the Southwest is on very large farms that use large equipment.   Wheat frequently is planted with 12 ft drills abreast and nearly always deep enough to reach moist soil and thus permit rapid germinaton.   Frequently soil packing equipment is used to prevent too rapid drying of the soil.   Call (1914), Martin (1926), Jardine (1916), and Leighty and Taylor (1927) have reported that early plowing and early seeding of winter wheat are usually beneficial, but early seeding exposes plants to streak-mosaic and there is, thus, a hazard involved.   There is considerable use of winter wheat for grazing (Anderson 1956).

Hard Red Spring Wheats

Nearly all the hard red spring wheat is grown in the northwest states of Minnesota, North and South Dakota, Montana, Idaho, and Washington.   In the drier areas of the western Dakotas, Idaho, and Washington, it is common to summer-fallow to conserve moisture. In the more humid areas, spring wheat is part of regular crop rotation practices.

It is a general custom to seed spring wheat with a drill as soon as the soil is sufficiently dry to work.   Early seeding provides a better opportunity to escape the effects of damage from hot weather, as well as insuring higher yield, as shown by Dungan and Burlison (1942).   Spring wheat predominates where winter wheats are killed by severe winters.

Durum Wheats

Amber and red durums are grown in a relatively small area in North and South Dakota and Minnesota.   The small quantity of red durum grown usually sells at a discount.   Seeding and harvesting practices for durum wheats are the same as for hard red springs.

Although there is a good demand for durum wheat in the United States, farmers are often reluctant to grow durum because of the higher seed requirements, rough awns, and generally weak straws that may lodge. In addition, the yield per acre of durum wheat until recently averaged less than spring and winter wheats. During the past 10 yr, yield per seeded acre for durum averaged 22.3 bu compared with 20.1 bu per acre for hard spring wheat and 23.5 bu for hard red winter.

Soft Red Winter Wheats

These are predominantly the wheats grown in the areas of greater rainfall, roughly east of the Mississippi River. The so-called "corn belt" also corresponds with the region of high production of soft red winter wheat. In fact, it is common to find farms where corn has been cut and winter wheat drilled in corn field stubble. Often wheat can be planted following corn by merely disking and harrowing the land. The use of commercial fertilizers to increase yields is commoner in the soft red winter region than other areas where wheat is grown. In general, soft red winter wheat is grown on smaller, diversified farms.

White Wheats

White wheats may be either spring or winter in habit. Principal areas of production are the western states of California, Idaho, Oregon, and Washington, and in Michigan in the Midwest, and New York in the East. In the United States, most white wheats are soft and have, in general, many of the same characteristics as soft red winter wheats. Cultural practices are about the same also.

HARVESTING WHEATS

Time of Harvesting

As shown in Table 1, somewhere in the world, wheat is being harvested every month in the year. In the United States, harvest starts in early May in southern Texas, reaches its peak in the hard red winter areas of the Southwest in early June, and ends in October in the northern portions of the red spring wheat area.

Methods of Harvesting

Not many years ago, the common method of harvesting grains was the binder, which still is used to a limited extent, especially on small farms. However, combines, small and large, have almost completely taken over wheat harvesting.   They cut and thresh the crop and are usually self-propelled.   Spring wheat is often windrowed or swathed before threshing.

From the combines, most spring and winter wheat is either stored on farms or taken by truck directly to local elevators where railway cars haul it to terminal markets.   Each year better highways and larger payloads increase the distances wheat can be hauled economically by truck.   Particularly in the "corn belt" area, soft red winter wheat is often stored on the farm, since smaller amounts are produced in that region.   Both farm storage and huge cooperative terminal storage facilities are common in the major wheat producing areas, where government programs subsidize both types of storage.

CLEANING GRAIN

Most grains, after harvest, are not cleaned either by farmers or by country elevators.   Wheat delivered at an elevator is sampled and graded with considerable accuracy and the elevator operator usually makes a price adjustment on the basis of moisture content, test weight, foreign material, and damaged or broken kernels.   Types of wheat kernel damage have been described in a Kansas Experiment Station Bulletin (1961).   It is not usual to clean or blend to regulate grade at small country elevator points; however, where facilities permit, it is often possible to improve grades by properly mixing the grain held in storage.   The important reason why wheat is not cleaned once it has been weighed and graded is that to discard any material whatsoever constitutes a weight loss.   Normally, whatever material is removed from wheat in the cleaning process has less value than wheat; consequently, cleaning can result in a direct financial loss. The task of cleaning wheat is left to the processor.

Wheat received at terminal elevators may be changed in a variety of ways such as by drying, washing, cleaning, separating, or sizing. Also, various grades and qualities can be collected and blended to supply processors wheat of uniform quality in large quantities.

STORAGE

Storage of cereal grains has been covered fully by Anderson and Alcock (1954).   Storage facilities vary from small inexpensive farm-type steel or wooden bins, to elaborate systems involving mechanized ventilation.   On farms, the circular type steel bin holding approximately 1,000 bu of grain is popular; while for terminal storage, a single elevator often has a capacity of several million bushels.

Marketing wheat commercially is based on the concept that, when reasonably dry, it can be stored for long periods without undergoing a change in grade. Maintaining large quantities of wheat in safe storage is a tremendously important economic consideration. Wheat is a biological material. It is alive and can deteriorate and die in storage if either the moisture content or temperature becomes too high. Wheat is also subject to attack by microorganisms, insects, and rodents when storage conditions are inadequate. For safe storage, the following must be considered: (1) moisture content, (2) temperature of the wheat, (3) infestation, and (4) type of bin.

Storage deterioration can be controlled by drying damp wheat to a safe moisture content and keeping it dry and cool, by providing good storage structures, and by aeration or forced ventilation. Fumigation or chemical protectants will control infestation. There is ample evidence that under normal conditions wheat does not deteriorate with storage age, at least within several years. Shellenberger *et al.* (1958) have shown that the length of storage of wheat has no appreciable influence on the storage properties of the flour milled from the wheat.

When wheat is stored improperly, deterioration is possible from many causes, including: (1) insect damage, (2) molds, (3) heat damage, (4) germ damage, (5) sprouting, (6) rotting, and (7) damage by rodents. Those factors not only can injure wheat and cause economic loss by lowering germination, but they also can cause actual loss in weight of wheat, bring about undesirable changes in odor, sanitary quality, and nutritive value, and can alter the chemical components such as proteins, lipids, and starch.

Experience, dating from antiquity, and modern research on grain storage have shown that moisture content and temperature are the principal influences in safe storage. The maximum moisture content at which wheat can be stored safely depends on the locality and the length of the storage period. Kelly *et al.* (1942) have shown that wheat may be safely stored for a year with a moisture content about 1% higher in North Dakota than in Kansas because of the lower average temperatures in North Dakota. For farm-type bins, wheat can be stored for a year or more at a moisture level of 13.0%. Hard wheats can often be kept safely with a moisture content of 14%.

UTILIZATION OF WHEAT IN THE UNITED STATES

Utilization as Human Food

The total US utilization of wheat, including food, seed, feed, and industrial use, has slightly increased year by year, and now amounts to over 640 million bushels annually. Per capita consumption of wheat flour has steadily declined from over 200 lb at the beginning of the century to 112 lb in 1967. However, population has increased from 76 million in 1900 to 200 million in 1967; consequently, total consumption of wheat flour increases.

The total quantity of wheat processed for the baking industry, including exports, amounted to 567,936 million bushels in 1967. Total flour production in 1967 was 250,789,000 cwt, including durum wheat flour which is used mostly in macaroni type products. The per capita consumption of macaroni, spaghetti, and noodles has steadily increased in recent years. Statistics on flour production and use are subject to constant change. The changes can be followed by consulting the most recent issues of Census of Manufacturers, The Wheat Situation, and Agricultural Statistics. Roughly, about 65% of the domestic consumption of flour now is utilized by the baking industry, 19% for family flour, including self-rising, 3% for durum products, and 13% in flour mixes.

Cereal grains have for centuries provided a main source of food for a large portion of the human race. Estimates are that between 35 and 40% of the total food energy of the people of the civilized world is derived from cereals. To promote growth and sustain normal life, the daily diets of man must contain adequate amounts and kinds of nutrients, including proteins, minerals, vitamins, plus the necessary carbohydrates to supply heat and energy. Wheat and wheat products have been able to supply these needs in an outstanding manner. An excellent review of the place of wheat in the human diet is contained in the bulletins published by the Wheat Flour Institute entitled, From Wheat to Flour (1965), and The Story of the Cereal Grains, General Mills, Inc. (1944).[1] Although the processing of wheat results in reducing total nutritive value of the processed product, enriching flour and baked products by adding vitamins and minerals retains the original nutritive values.

Waggle et al. (1967) have provided extensive analyses of several classes of wheat and milled products including flour, shorts, bran, red dog, and germ. The results include an analysis of 17 amino acids, 9 vitamins, 15 minerals, and gross energy content. The amino acid content of four classes of wheat is shown in Table 5. The most limiting amino acid in the protein of white flour is lysine, but Hegsted (1965) suggests that the ingestion of 228 gm of white flour per day would satisfy the deficiency. Other amino acid requirements are listed in Table 6.

---

[1] Wheat Flour Institute, 14 East Jackson Boulevard, Chicago, Ill. General Mills, Inc., 9200 Wayzata Boulevard, Minneapolis, Minn.

TABLE 5

AMINO ACID COMPOSITION OF WHEAT SAMPLES[1]

	HRW (%)					HRS (%)		White (%)	SRW (%)
	9001	9002	9008	9007	9010	9009	9003	9005	9006
Protein	10.5	13.1	11.7	11.4	11.8	10.2	13.5	9.9	12.0
Amino acids									
Lysine	0.29	0.35	0.34	0.34	0.35	0.28	0.34	0.32	0.35
Histidine	0.23	0.30	0.30	0.29	0.30	0.24	0.27	0.26	0.31
Arginine	0.49	0.60	0.61	0.62	0.63	0.50	0.54	0.55	0.63
Aspartic acid	0.56	0.63	0.67	0.66	0.65	0.55	0.69	0.57	0.65
Threonine	0.34	0.40	0.35	0.35	0.36	0.33	0.40	0.32	0.38
Serine	0.59	0.69	0.54	0.54	0.57	0.53	0.69	0.52	0.63
Glutamic acid	3.58	4.46	4.16	3.93	3.85	3.72	4.82	3.45	4.27
Proline	1.11	1.54	1.36	1.17	1.28	1.15	1.55	1.06	1.36
Glycine	0.44	0.56	0.56	0.52	0.55	0.46	0.59	0.47	0.53
Alanine	0.39	0.48	0.48	0.43	0.47	0.40	0.50	0.40	0.48
Cystine	0.30	0.38	0.32	0.30	0.33	0.31	0.34	0.29	0.35
Valine	0.41	0.56	0.57	0.59	0.59	0.50	0.56	0.48	0.56
Methionine	0.19	0.24	0.19	0.17	0.20	0.18	0.22	0.16	0.21
Isoleucine	0.34	0.46	0.50	0.47	0.48	0.38	0.49	0.38	0.44
Leucine	0.73	0.94	0.89	0.83	0.86	0.74	0.97	0.74	0.88
Tyrosine	0.34	0.43	0.39	0.38	0.38	0.33	0.42	0.32	0.38
Phenylalanine	0.50	0.67	0.61	0.58	0.60	0.51	0.67	0.49	0.62

[1] All values reported on 14% moisture basis.

TABLE 6

ESTIMATED DAILY AMINO ACID REQUIREMENTS AND
AMOUNT OF WHEAT OR WHEAT FLOUR REQUIRED

Amino Acid	Average Requirement (Mg/day)	Quantity Needed to Meet Requirement (Gm)	
		Whole Wheat	White Flour
Tryptophan	170	104	130
Phenylalanine	260	40	45
Threonine	375	68	86
Isoleucine	550	95	114
Lysine	545	150	228
Methionine	200	99	145
Total S	700	140	200
Valine	620	100	137
Leucine	725	81	90

Wheat and wheat products contribute substantially to the world's food supply by providing energy, protein that contains a good assortment of amino acids, fats, minerals, and vitamins. Because of these facts and the problem of malnutrition in the world which is discussed widely in scientific, industrial, and governmental circles, a great deal of work has been done to develop wheat-based products for world use. This subject and the developments sponsored by The Millers' National Federation and the US Dept. of Agriculture are discussed by Sullivan (1967) and by Senti *et al.* (1967)

Bulgur, Wurld Wheat, rolled wheat, and a wheat-based beverage are products for which markets are being developed in countries where serious malnutrition exists. In the normal flour milling process about 25% of the total wheat is sold for animal feed. It has long been recognized that the aleurone and outer layers of the wheat kernel have superior nutritive value because of the quantity and quality of protein, vitamin, and mineral content, and that they have not been used for human food. Methods to make such products of high nutritive value available in acceptable forms for food are being investigated.

Feed and Industrial Uses

Traditionally, wheat has been considered a good livestock and poultry feed. It compares favorably with corn in feed value and is considered superior to barley. Although government subsidy programs initiated in 1947 often have increased the price of wheat relative to feed grains, the marketing certificate programs introduced in 1964 adjusted prices and now the amounts of wheat fed on the farm and used in mixed feed and processed feed have steadily increased. For example, during 1959–63 the average amount of wheat feed was 30.3 million bushels, while the average for 1964–66 was 94.3 million bushels, and is estimated to be 125–150 million bushels for 1967.

When wheat is processed, portions of the kernel not used for food go into feed use as bran, shorts, red dog, germ meal, and mixed feed. The nutritional attributes of such products have been studied extensively (National Research Council 1959) but one big difficulty with the use of flour mill by-products has been that they vary so widely in composition. Wheat, for example, can vary in protein content from 7 to 18%. Obviously the bran, shorts, and other products from wheat of different composition also differ. Lack of standardization has affected use of mill feed by the formula feed industry, and consequently, studies were instigated by Farrell *et al.* (1967), Waggle *et al.* (1967), and others and the results were published in The Millers' National Federation Millfeed Manual (1967). Analytical and nutritional information was collected on mill feed from hard spring and winter wheat, soft red winter, and white wheats. Much progress has been made in providing basic information on millfeeds of the feed industry.

The wheat plant also provides animal feed in the form of hay and pasture. Winter wheat in many years is an excellent source of pasture in the fall and early spring for cattle, particularly.

During latter stages of maturation of the wheat kernel, there is evidence of a decrease in the concentration of several of the more nutritionally important amino acids.   Lambert *et al.* (1968) investigated the possibility that preripe wheat might have more nutritive value than fully matured grain because of the greater availability of the amino acids.   Tests with chicks and rats failed to show any nutritional benefits from feeding preripe wheat.

Only relatively small amounts of wheat and wheat flour are consumed industrially.   Government farm price support programs drastically limit industrial uses of wheat by the economics of the situation.   Wheat can be used to manufacture industrial alcohol, gluten, monosodium glutamate, starch, pastes, and core binder. There is a market for a limited amount of wheat malt which is used chiefly to enhance the diastatic activity of wheat flour for baking. Wheat malt has only limited use in the brewing and distilling industries compared with barley malt, because of price.   Usually clear and low-grade flours are used in the manufacture of wheat starch, monosodium glutamate, and gluten or pastes for plywood adhesives, book binding, or paper hanging.   Wheat products have been used to some extent in foundries as core binder in making molds for castings.

There is a large and expanding market for cereal starches, but wheat starch has played only a minor part in this development over the past 50 yr.   Processes for converting starch to dextrose and sugar syrups have improved greatly.   Starches are now modified to alter their properties for special purposes in the paper, textile, adhesive, and chemical industries.   Whenever wheat is surplus and less expensive than other grains, the industrial market is available.

IMPORTANCE OF WHEAT TO THE ECONOMY OF THE UNITED STATES

The great adaptability of wheat to varying climates and soils is evident from wheat growing in sufficient quantities to be reported in 42 of the 50 states.   Wheat is unimportant only in the New England states.   The principal wheat producing states are Kansas, North Dakota, Oklahoma, Texas, Montana, Nebraska, Colorado, and Washington.   Present wheat production exceeds 1.5 billion bushels per year, which has cash value to farmers of approximately $2 billion a year.   It is an important source of farm income.

Transporting, handling, financing, storing, and processing wheat involve a cross section of US business activities.   Exporting wheat from the United States provides an important source of foreign exchange.   Wheat is often used to help provide economic and political stability in other countries, and is an instrument in foreign-policy making.   Of all cereal grains, wheat is the one that enters most into international trade.

COMPOSITION OF WHEAT

The composition of wheat varies greatly from area to area and from year to year within a given area.   The possible range of variability of composition of wheat in the United States in a crop year is indicated in Table 7.

TABLE 7

APPROXIMATE COMPOSITION OF WHEAT

Determination	Composition Range (%)	
	Low	High
Protein (N × 5.7)	7.0	18.0
Mineral matter (ash)	1.5	2.0
Lipids (fat)	1.5	2.0
Starch	60.0	68.0
Cellulose (fiber)	2.0	2.5
Moisture	8.0	18.0

Hard red spring or winter wheat usually has a protein content of at least 12%.   Soft red or white wheats for cake and cookie flour manufacturing are weak with protein contents from 8 to 10%.   Wheats that are between the bread and pastry types are used for cracker, doughnut, and all-purpose flours.

In addition to the approximate composition of wheat, which includes only the broad chemical constituents, a vast amount of analytical data has been assembled on the amino acid content of the wheat proteins, the composition of the mineral matter, enzyme components, vitamin content, and the properties of the starch and other carbohydrate material.   Bailey (1944) has presented the significant facts and data relating to the substances present in wheat.   More recent information on this subject has been covered by Jacobs (1951), Hlynka (1964), Farrell et al. (1967), and Waggle et al. (1967).   The proteins, carbohydrates, lipids, enzymes, and pigments of wheat are fully covered in the book edited by Hlynka (1964) referred to previously.

QUALITY TESTS

Physical Tests Applied to Wheat

Miller and Johnson (1954), Shellenberger and Ward (1967), and Finney and Yamazaki (1967) have reviewed tests to evaluate wheat quality. The following physical tests are commonly applied to wheat: (1) federal grade, including test weight; (2) kernel hardness; (3) gluten washing; (4) internal infestation evaluation; (5) density; (6) thousand-kernel weight; (7) pearling index; and (8) granulation and particle size.

Methods for performing the physical tests are included in the hand book of the Official Grain Standards of the United States (Anon. 1964) and Cereal Laboratory Methods (Anon. 1962).

Official grades of wheat include seven categories, namely, Hard Red Spring, Durum, Red Durum, Hard Red Winter, Soft Red Winter, White, and Mixed Wheat. Within each class, subclasses and numerical grades are determined by test weight, percentage of hard and vitreous kernels, extent of damaged kernels, and amounts of foreign material and wheats of other classes. Those factors are summarized in Table 8, using Hard Red Winter as an example. Determining the grade is the first step in appraising wheat quality and is the dominant consideration in marketing and handling wheat. Moisture and protein content information usually are considered along with the grade and also the extent of infestation, especially when wheat is evaluated for processing.

TABLE 8

UNITED STATES GRAIN STANDARDS FOR HARD RED WINTER WHEAT[1]

Grade	Minimum Test Weight per Bushel (Lb)	Heat Dam-age	Maximum Limits of Defects (%)				
			Dam-aged Ker-nels	For-eign Mate-rial	Shrunken and Broken Kernels	Defects (Total)	Wheat of Other Classes
1	60.0	0.1	2.0	0.5	3.0	3.0	3.0
2	58.0	0.2	4.0	1.0	5.0	5.0	5.0
3	56.0	0.5	7.0	2.0	8.0	8.0	10.0
4	54.0	1.0	10.0	3.0	12.0	12.0	10.0
5	51.0	3.0	15.0	5.0	20.0	20.0	10.0
Sample grade							

[1] There are three subclasses: (a) Dark Hard Winter, (b) Hard Winter, and (c) Yellow Hard Winter, based on per cent of dark, hard, and vitreous kernels.

Milling quality is a broad term used to embody the many factors that affect the milling process. Among them are the response of the wheat to conditioning, the millability or reduction of stocks, and the flour yield. Many attempts have been made to standardize and evaluate experimental milling procedures but progress has been slow. Shellenberger and Ward (1967) have reviewed the literature on experimental milling. Hard, soft, and durum wheats respond in vastly different ways to milling operations.

The gluten washing test is a physical determination of quality once applied generally to wheat in the United States, but now replaced almost entirely by the protein test. The gluten test is, however, used in many places in the world. Essentially the test involves removing starch from a flour-water dough and collecting a cohesive mass of gluten, which can be evaluated both as to quantity and quality. Weaknesses of the gluten test are lack of reproducibility and differences in interpreting gluten quality.

Chemical Tests Applied to Wheat

Moisture, ash, and protein tests are the most widely used on wheat. The Kjeldahl protein test normally is made on all hard wheat reaching terminal markets but it is not an official part of US Grain Standards. It provides supplementary information to that given on the grade certificate. To report protein content on a uniform moisture basis, the moisture test is necessary. Miller and Johnson (1954) state that moisture is related to quality of wheat and flour in at least three ways: (1) flour yield varies inversely with moisture content; (2) composition percentages are inversely related to percentages of moisture present; and (3) deterioration of grain during storage may depend on the moisture relationships in the wheat kernel.

The ash test is significant as a way to determine flour grade. Since the ash content of bran is about 20 times that of endosperm, the ash test indicates how thoroughly bran and germ were separated from the kernel endosperm. There is considerable variation in the amount of mineral matter in wheat, depending on the class of wheat and the area it was grown; therefore, the ash test is often applied to wheat as well as to flour.

Color tests, such as the Pekar or the slick test, also are used to judge milling results or flour grade, and recently photoelectric methods such as the Agtron (Gilles 1963) for assessing flour color have come into use.

Many attempts have been made to appraise wheat quality on the basis of the amino acid composition of wheat gluten. Determining the amino acid composition of wheat is no longer a tedious and expensive procedure, but Pence et al. (1950) failed to show essential differences in the amino acid content of 17 flours milled from different varieties and types of wheat. Recent investigations on the amino acids of wheat were discussed previously.

Physical-chemical Tests

Starch damage is a subject of recent concern to wheat processors and to the baker. The extent of starch damage alters flour characteristics and end use.

The amylase activity of wheat and flour are measured by the Falling Number test, Blish-Sandstedt method, and the amylograph. Descriptions of those tests and others mentioned can be found in *Cereal Laboratory Methods* (1962) and *Wheat and Wheat Technology* (Hlynka 1964).

When wheat deteriorates during storage, various chemical changes occur. Under normal storage conditions the changes occur slowly, but under unfavorable storage conditions the changes can be rapid. Christensen and Kaufmann (1964) have shown that fungi are involved in all cases of grain spoilage. Among the chemical changes associated with wheat deterioration, the breakdown of fats by lipases to liberate free fatty acids has been used as a criterion of soundness. Pomeranz and Shellenberger (1966) concluded that fat acidity does not provide valid information to evaluate the baking quality of stored wheat. High fat acidity may or may not be indicative of impaired baking quality of wheat.

A large number of tests based on the imbibitional properties of the wheat proteins have been developed and used. Berliner and Koopman (1929) evolved a method to measure the swelling power of gluten. Gortner (1924) and his co-workers did much to develop and popularize the use of the viscosity test as a measure of flour quality. Zeleny (1947) presented a simple sedimentation test, which has been proposed as a method to estimate the baking quality of wheat. The method consists essentially of suspending flour particles in a graduated cylinder containing dilute lactic acid. The rate of sedimentation is a measure of the hydration capacity of the flour proteins and is an index of quality. The sedimentation test applied to approximately 360 cargos of wheat in world commerce by Shellenberger (1958) showed a correlation of 0.726 between sedimentation values and the overall quality score of the wheats. This was a significantly higher correlation than that between protein content and quality score.

Many investigators have attempted to use the heats of hydration of wheat flour as an index of quality. The method thus far has not proved practical since the differences in commercial values between flours of known different properties are slight. Peptization and gluten fractionation methods as a means of measuring gluten quality also have received much attention. Gas production and gas retention measurements in dough have been developed to measure those properties automatically. The bread baking test is a practical way to measure gas production and retention.

Varietal Differences in Gluten Quality

Many techniques have been developed to separate gluten from the other constituents of wheat flour. The washing-out process can be done by hand or mechanically. A quantitative test consists merely of weighing the gluten either wet or after being dried, but estimating gluten quality is a more difficult matter.

The measurement of gluten quality has been based on four general principles, namely: expansion by heat, recovery from compression, gluten extension, and gluten relaxation. The more satisfactory methods for appraising quality are those that use the extension or relaxation of gluten.

Although there is agreement that many properties of dough are due to the gluten component of flour, most cereal chemists in the United States prefer to study the properties of doughs rather than to separate gluten and thus attempt to appraise quality. Regardless of the methods used, differences in gluten properties between classes and varieties of wheat are obvious.

The meaning of quantity of protein in wheat or flour is readily understood since determining protein is a standardized procedure. The meaning of gluten quality is not parallel because quality is fixed by the characteristics inherent in the wheat kernel and by the purposes for which the flour is to be used. There are three well-defined gluten ranges: strong hard wheat gluten, weak soft wheat gluten, and durum type gluten; the last is poor for bread dough but ideal for macaroni products.

Varieties within the classes of hard red spring, hard red winter, or hard white wheat can possess wide differences in gluten quality. To be of top quality, a variety of hard wheat should possess high dough absorption, have wide mixing and fermentation tolerance, and be capable of producing bread of good volume, grain, and texture. Some wheat varieties possess those necessary attributes to a high decree while others do not. The selection of satisfactory wheat for specific purposes depends upon the determination of the necessary gluten quality.

Quality Tests for Milling Wheat

Wheat is marketed on the basis of the Federal Grade. The dominant factor affecting grade, after the class has been established, is test weight. If wheat is designated as "tough," it means that it has a moisture content between 14 and 15.5% if it is soft wheat, and between 14.5 and 16% if it is hard wheat. In addition to test weight, determining grade is based on the extent of damaged kernels and the amount of foreign material and wheats of other classes. All of those factors constitute an effective means to describe wheat for marketing purposes; however, a great deal more needs to be known when it is purchased to manufacture specific types of flour.

The grade of a wheat indicates its freedom from foreign materials and the probable cleaning loss involved in preparing it for milling. At lower test weights, especially, there is some relation between flour yield and test weight. Therefore, the first appraisal of wheat quality for processing is obtained from the official grade but tests for protein, moisture, and ash are commonly applied to assist in the evaluation of wheat for processing purposes. The protein test is the best single test that can be applied to estimate quality. The protein test is, however, far from satisfactory, because no one test can be expected to appraise the quality of a complex biochemical system like the one in a wheat kernel. Yet, the correlation between protein content and quality is high. Low protein content in wheat indicates quality for cake and pastry flour, while high protein points to bread flour use.

The moisture determination is useful in considering storage and conditioning requirements. The ash determination need not necessarily be considered a part of the essential quality information. If wheat is purchased from an area where the mineral content in the kernel is unusually high, then it is advisable to have a record of the wheat ash.

To obtain a satisfactory insight into wheat quality, it is necessary to mill the wheat and subject the flour obtained to tests not only of protein quality but also of baking strength. The Brabender farinograph is one of the most commonly used instruments to measure absorption and physical dough properties of wheats, and Brabender (1932, 1965) has discussed the procedures for these tests. This instrument measures plasticity and mobility of dough when subjected to mixing at a constant temperature. Although the farinograph is useful to estimate absorption and dough development properties of both hard and soft wheat flour, it has greater utility for hard wheat. The viscosity test (Bayfield 1934) is used extensively to appraise soft wheat quality.

Wheats destined for milling are often subjected to indirect measurements for amylase activity by means of gas production determinations, Falling Number values, or use of the amylograph (Anon. 1962). Most hard wheat flours require the adjustment of the amylase level, but wheats unusually high or low in amylase activity require special consideration for use in mill blends.

In North America it is common for the milling industry to have an accurate estimation of the baking quality of the wheats used to make up the mix. Standard type tests have been developed to help appraise quality of bread, cakes, and pastry products. The experimental bread baking test has reached a high state of acceptance as a measure of wheat quality. The real appraisal of wheat quality is based on its use and Finney and Yamazaki (1967) have provided a good review of this subject.

BIBLIOGRAPHY

ANDERSON, J. A., and ALCOCK, A. W. 1954. Storage of Cereal Grains and Their Products. Am. Assoc. Cereal Chemists, St. Paul, Minn.

ANDERSON, K. L. 1956. Winter wheat pasture in Kansas. Kansas Agr. Expt. Sta. Bull. *345*.

ANON. 1953. Encyclopedia Britannica, Encyclopedia Britannica, Inc., Chicago, Ill.

ANON. 1956. Wheat varieties commercially important in the hard red winter wheat area. Kansas State Univ. Ext. Serv. Ext. Circ. *254*.

ANON. 1959. Natl. Academy Sci.—Natl. Res. Council, Committee on animal nutrition and national committee on animal nutrition in Canada. Joint US-Canada Tables of Feed Composition. Publ. *659*.

ANON. 1962. Cereal Laboratory Methods. Am. Assoc. Cereal Chemists, Inc., St. Paul, Minn.

ANON. 1964. Official Grain Standards of the United States. US Dept. of Agr., Agr. Marketing Serv., Grain Div., Washington, D.C.

ANON. 1967. Millfeed Manual. The Millers' National Federation, Chicago, Ill.

BAILEY, C. H. 1944. The Constituents of Wheat and Wheat Products. Reinhold Publishing Corp., New York.

BAYFIELD, E. G. 1934. Soft wheat studies. II. Evaluating experimentally milled flours with the aid of viscosity, fermentation and baking tests. Cereal Chem. *11*, 121–140.

BERLINER, E., and KOOPMAN, J. 1929. The determination of gluten in wheat flour. Z. ges. Muhlenw. *6*, 57–63, 75–82, 91–93.

BRABENDER, C. W. 1932. Studies with the farinograph for predicting the most suitable types of American export wheats and flours for mixing with European soft wheat flours. Cereal Chem. *9*, 617–627.

BRABENDER, C. W. 1965. Physical dough testing, past, present and future. Cereal Sci. Today *10*, 291–304.

BRADBURY, D., CULL, I. M., and MACMASTERS, M. M. 1956. Structure of the mature wheat kernel. I. Gross anatomy and relationship of parts. Cereal Chem. *33*, 329–342.

BRADBURY, D., MACMASTERS, M. M., and CULL, I. M. 1956A. Structure of the mature wheat kernel. II. Microscopic structure of pericarp, seed coat, and other coverings of the endosperm and germ of hard red winter wheat. Cereal Chem. *33*, 324–360.

BRADBURY, D., MACMASTERS, M. M., and CULL, I. M. 1956B. Structure of the mature wheat kernel. III. Microscopic structure of the endosperm of hard red winter wheat. Cereal Chem. *33*, 361–373.

CALL, L. E. 1914. The effect of different methods of preparing a seed bed for winter wheat upon yield, soil moisture and nitrates. J. Am. Soc. Agron. *6*, 249–259.

CHRISTENSEN, C. M., and KAUFMANN, H. H. 1964. Questions and answers concerning spoilage of stored grains by storage fungi. Agr. Ext. Serv., Univ. of Minnesota and USDA.

DAFTARY, R. D., and POMERANZ, Y. 1965. Changes in lipid composition in maturing wheat. J. Food Sci. *30*, 577–582.

DUNGAN, G. H., and BURLISON, W. L. 1942. Spring wheat: adaptability for Illinois. Ill. Agr. Expt. Sta. Bull. *483*.

FARRELL, E. P., WARD, A., MILLER, G. D., and LOVETT, L. A. 1967. Extensive analyses of flours and millfeeds made from nine different wheat mixes. I. Amounts and analyses. Cereal Chem. *44*, 39–47.

FINNEY, K. F. 1954. Contributions of the Hard Winter Wheat Quality Laboratory to wheat quality research. Trans. Am. Assoc. Cereal Chemists *12*, 127–142.

FINNEY, K. F., and YAMAZAKI, W. T. 1967. Quality of hard, soft and durum wheats. *In* Wheat and Wheat Improvement. Am. Soc. Agronomy, Madison, Wis.

FUKASAWA, H. 1958. Fertility restoration of cytoplasmic male-sterile emmer wheats. Wheat Information Serv. *7*, 21.

GILLES, J. 1963. The Agtron. Cereal Sci. Today *8*, 40–42.

GORTNER, R. A. 1924. Viscosity as a measure of gluten quality. Cereal Chem. *1*, 75–81.

HEGSTED, D. M. 1965. Wheat: challenge to nutritionists. Cereal Sci. Today *10*, 257–259.

HLYNKA, I. 1964. Wheat Chemistry and Technology. American Association of Cereal Chemists, Inc., St. Paul Minn.

HOSENEY, R. C., and FINNEY, K. F. 1967. Free amino acid composition of flours milled from wheats harvested at various stages of maturity. Crop Sci. *7*, 3–5.

HOSENEY, R. C., FINNEY, K. F., SHOGREN, M. D., and POMERANZ, Y. 1968A. The role of flour-solubles in breadmaking. Cereal Chem. (in press).

HOSENEY, R. C., FINNEY, K. F., SHOGREN, M. D., and POMERANZ, Y. 1968B. Characterization of gluten protein groups by starch-gel electrophoresis and bread-baking. Cereal Chem. (in press).

JACOBS, M. B. 1951. The Chemistry and Technology of Food and Food Products. Interscience Publishers, New York.

JARDINE, W. M. 1916. Effect of rate and date of sowing on yield of winter wheat. J. Am. Soc. Agron. *8*, 163–166.

JENNINGS, A. C., and MORTON, R. K. 1963A. Changes in carbohydrates, protein and non-protein nitrogenous compounds of developing wheat grain. Australian J. Biol. Sci. *16*, 318–331.

JENNINGS, A. C., and MORTON, R. K. 1963B. Amino acids and protein synthesis in developing wheat endosperm. Australian J. Biol. Sci. *16*, 384–394.

JENNINGS, A. C., and MORTON, R. K. 1963C. Changes in nucleic acids and other phosphorus-containing compounds of developing wheat grain. Australian J. Biol. Sci. *16*, 332–341.

KELLY, C. F., STAHL, B. M., SALMON, S. C., and BLACK, R. H. 1942. Wheat storage in experimental farm-type bins. US Dept. Agr. Circ. *637.*

KIHARA, H. 1961. Cytoplasmic male sterility of common wheat. Ann. Rept. Nat. Institute Gen. No. *12*, Misima, Japan.

LAMBERT, M. A., DEYOE, C. W., SHELLENBERGER, J. A., and SANFORD, P. E. 1968. The feeding value of preripe wheat. Feedstuffs (in press).

LEIGHTY, C. E., and TAYLOR, J. W. 1927. Rate and date of seeding and seed-bed preparation for winter wheat at Arlington Experiment Farm. US Dept. Agr. Tech. Bull. *38.*

MARTIN, J. H. 1926. Factors influencing results from rate and date of seeding experiments with wheat in the western United States. J. Am. Soc. Agron. *18*, 193–225.

MILLER, B. S., and JOHNSON, J. A. 1954. A review of methods for determining the quality of wheat and flour for breadmaking. Kansas Agr. Expt. Sta. Tech. Bull. *76.*

PENCE, J. W. *et al.* 1950. Characterization of wheat gluten. II. Amino acid composition. Cereal Chem. *27*, 335–341.

PERCIVAL, J. 1921. The Wheat Plant. E. P. Dutton and Co., New York.

POMERANZ, Y. 1967. Wheat flour lipids. Bakers Dig. *41*, 48–50.

POMERANZ, Y., FINNEY, K. F., and HOSENEY, R. C. 1966. Amino-acid composition of maturing wheat. J. Sci. Food Agr. *17*, 485–487.

POMERANZ, Y., and SHELLENBERGER, J. A. 1966. The significance of free fatty acids in cereals. Am. Miller and Processor *94*, 9–11.

POOL, R. J. 1948. Marching with the Grasses. Univ. Nebr. Press, Lincoln, Nebr.

QUISENBERRY, K. S., and REITZ, L. P. 1967. Wheat and Wheat Improvement. American Society of Agronomy, Madison, Wis.

SANDSTEDT, R. M. 1965. Fifty years of progress in starch chemistry. Cereal Sci. Today *10*, 305–358.

SCOTT, G. E. 1955. The development of the wheat kernel. Unpublished Thesis. Kansas State Univ., Manhattan.

SCOTT, G. E., HEYNE, E. G., and FINNEY, K. F. 1957. Development of the hard red winter wheat kernel in relation to yield, test weight, kernel weight, moisture content and milling and baking quality Agron. J. *49*, 509–513.

SENTI, F. R., COPLEY, M. J., and PENCE, J. W. 1967. Protein-fortified grain products for world uses. Cereal Sci. Today *12*, 426–430.

SHELLENBERGER, J. A. 1957. The story of wheat development in Kansas. Cereal Sci. Today 2, 74–78.

SHELLENBERGER, J. A. 1958. Survey of the quality of European wheat imports. Kansas Agr. Expt. Sta. Bull. 396.

SHELLENBERGER, J. A., MILLER, D., FARRELL, E. P., and MILNER, M. 1958. Effect of wheat age on storage properties of flour. Food Technol. 12, 213–221.

SHELLENBERGER, J. A., and WARD, A. B. 1967. Experimental milling. In Wheat and Wheat Improvement. Am. Soc. Agronomy, Madison, Wis.

STEWART, W. D. P. 1967. Nitrogen-fixing plants. Sci. 158, 1426–1432.

STORCK, J., and TEAGUE, W. D. 1952. Flour for Man's Bread. Univ. of Minn. Press, Minneapolis.

SULLIVAN, B. 1965. Wheat protein research. Fifty years of progress. Cereal Sci. Today 10, 338–361.

SULLIVAN, B. 1967. Wheat-based products for world use. Cereal Sci. Today 12, 446–448.

WAGGLE, D. H. et al. 1967. Extensive analyses of flours and millfeeds made from nine different wheat mixes. II. Amino acids, minerals, vitamins and gross energy. Cereal Chem. 44, 48–60.

WALL, J. S. 1967. Origin and behavior of flour proteins. Bakers' Dig. 41, 36–44.

WHITEHAIR, N. V., and ENIX, J. R. 1961. Wheat kernel damage. Kans. Agr. Expt. Sta. Bull. 19.

WILSON, J. A., and ROSS, W. M. 1961. Cross-breeding in wheat, Triticum aestivum. I. Frequency of pollen-restoring character in hybrid wheat having Aegilops ovata cytoplasm. Crop. Sci. 1, 191–193.

ZELENY, L. 1947. A simple sedimentation test for estimating the bread-baking and gluten qualities of wheat flour. Cereal Chem. 24, 465–475

CHAPTER 2

Samuel A. Matz | Corn

INTRODUCTION

Origin

Corn (*Zea mays* L.) originated in the Western hemisphere. It was the only cereal systematically cultivated by the American Indian although some other grains were harvested from the wild state. Columbus found corn being cultivated on Haiti, where it was called mahiz. From this Arawak Indian word was derived the name maize that is used in Europe to distinguish the cereal from other grains which are called "corn" (Hunt 1915).

The tribe Tripsaceae to which corn belongs differs quite widely from the tribe Hordeae, to which the other common cereal grains (wheat, rye, and barley) belong. In the same tribe with corn are teosinte (*Euchlaena mexicana* Schrad.), a subtropical plant regarded by some as the ancestor of corn, gama grass (*Tripsacum dactyloides* L.), and adlay or Job's tears (*Coix lacryma-Jobi*).

Since the grain-bearing part of corn is inclosed in tenacious leaf-sheaths, it is incapable of reseeding itself. Wild forms do not exist. The evolutionary path leading to modern corn is not clear, but recent archeological findings have partially explained its development. Mangelsdorf *et al.* (1964) described an evolutionary sequence based on about 24,000 maize specimens found in 5 caves in the Tehuacán Valley of southern Puebla and northern Oaxaca, Mexico. Some of these specimens were nearly 7,200 yr old. It was deduced that wild corn probably had a 1-in. ear with 2 husks borne high on the stalk. At maturity, the husks opened to permit dispersal of the seeds. Above the ear was a male spikelet, about 1 to 2 in. in length. The kernels were rounded and either brown or orange. Repeated selection of seed from plants with larger kernels, more kernels on each cob, and other desirable features led to corn as we know it.

It is apparent that this wild corn no longer exists. Either it was completely displaced by domesticated corn which was sowed in all places suitable for the plant, or cultivated corn hybridized with the wild corn rendering the latter incapable of independent propagation.

Corn can be crossed with teosinte and some of the hybrids are

39

Courtesy of US Dept. Agr.

FIG. 10. *Euchlaena Mexicana*

Plant, much reduced;   pistillate inflorescence enclosed in
bract (a) and with portion removed (b); lateral view of rachis
joint and fertile spikelet (c), and dorsal view of same, showing
first glume (d).   Not to scale.

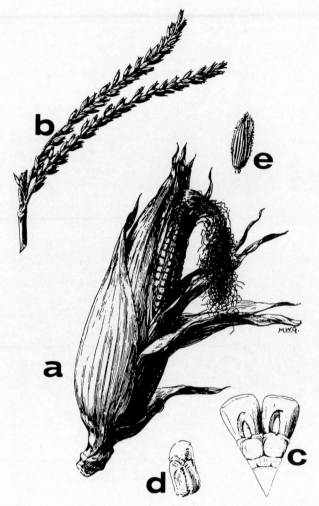

Courtesy of US Dept. Agr.

FIG. 11.   *Zea Mays*

Pistillate inflorescence (a) or ear, and two branches of
staminate inflorescence (b) or tassel;   pair of pistillate
spikelets attached to rachis (cob) with mature grains (c),
the second glume showing;   single pistillate spikelet soon
after flowering (d);   and staminate spikelet (e).   Not to
scale.

even fairly fertile. These "tripsacoids" apparently became the most common maize in the period from about 1000 to 200 BC, according to the Tehuacán valley specimens. The tripsacoid complex eventually gave rise to Nal-Tel and Chapalate, two Mexican corn races which still exist, and to another type called late tripsacoid.

Over the several thousand years of its evolution under cultivation there has been no change in the fundamental botanical characteristics of the corn plant. That is, the changes which have occurred were in the size of the parts of the plant and in productivity.

Some of the earlier theories on the origin of corn were well summarized by Wallace and Brown (1956). Weatherwax (1954) and some other authorities favored the hypothesis that corn, teosinte, and tripsacum evolved from a single ancestral stock by ordinary divergent evolution. Other botanists, on the basis of genetic evidence, consider *Zea* to be older than teosinte, the latter having arisen from a relatively recent cross of corn with gama grass.

Types of Corn

By far the greatest production of corn is field corn of the dent and flint types. When the unqualified term "corn" is used this kind of corn is meant. Sweet corn and popcorn are also of economic importance.

Sweet corn differs from field corn in that a large proportion of the carbohydrates of the kernel is present as glucose polymers of fairly low molecular weight (dextrins) rather than as starch granules. As a consequence, the kernels of sweet corn retain their tender and succulent texture, and their sweet taste for a longer period of time during their development. Sweet corn kernels, when matured and dried, are as hard as those of field corn, although they have a wrinkled surface. On the other hand, certain varieties of field corn are often sold as sweet corn when in the immature stages. This is particularly true in the southern United States where the greater susceptibility of sweet corn to disease and insect infestation has prohibited its economical production until recently.

Sweet corn has been regarded by some botanists as a distinct species or subspecies (*Zea mays saccharata*) extant since prehistoric times, while other authorities (Erwin 1951) consider it to be a field-corn mutation of fairly recent origin. The latter view is the one which prevails today. The mutated gene, su_1, is presumably one of those responsible for initiating the mechanism whereby the carbohydrates of the kernel are transformed into granules containing high molecular weight branched starches. There are other gene combinations that give sweet corn, but the common commercial type is due to su_1. Single kernels having the characteristics of sweet corn can be observed on ears of flint corn grown under controlled pollination. It is noteworthy that sweet corn differs from field corn in several respects. Sweet corn is more susceptible to attacks by pathogens, for example. These additional differences could result from the operation of the same mutant gene which interferes with starch granule production or from linked genes controlling susceptibility.

Another type of sweet corn called Supersweet has been introduced by the University of Illinois. It is genetically homozygous for a mutant gene called sh_1, which blocks all starch synthesis. The kernels contain 35% sucrose at milk stage and maturity. Currently, Supersweet is being grown only in home gardens, but it may soon meet some demand for commercial use.

Seed-bed preparation, planting, and cultivation of sweet corn are similar to the procedures used for field corn. Harvesting is usually performed by machines which cut the ears off the stalk so that bruising of the tender kernels is kept to a minimum. The time of harvesting is quite critical and is usually about 18 days after silking. Even the most tolerant varieties retain satisfactory canning quality for a period of only 6 to 8 days. In other varieties, this period of optimum quality may be as short as two days. Sweet corn yields from about $1/2$ to $2/3$ as much food product dry matter per acre as do the best varieties of field corn under similar conditions. Most aspects of sweet corn growing and processing have been excellently presented by Huelsen (1954).

Sweet corn is marketed fresh, frozen, or canned. A very small proportion is also dehydrated by methods conducive to rehydration.

Most canned corn is marketed either as cream style or as whole kernel, although a small amount of corn-on-the-cob and some combination items (e.g., corn with peppers) are canned. Cream style corn includes portions of kernels cut from the cob and the scraped residue of the portions remaining on the cob. Small amounts of corn or wheat starch, natural or modified, may be added to thicken the can contents.

All, or nearly all, canned corn is sweet corn. The amount of field corn canned (as such, excluding hominy) is negligible. The corn for canning is grown under contract with a processor. The canner provides the farmer with treated seed and designates the date of planting. On maturity, the corn is purchased at a price which may be fixed at the time of the signing of the contract or may be related by some factor to the price of field corn at the time of delivery. The contractual arrangement is necessary because the factory must have a constant flow of raw material over an extended canning period, and the farmer must have an assured outlet for this very perishable crop.

About $4^1/_2$ billion pounds of sweet corn were harvested from about 580,000 acres in 1965.

Corn is received at the cannery with the husks on the ears. The corn is fed from the trucks into chutes leading to machines which cut off the ends of the cobs and strip the freed husk from the ears. Washing by high pressure jets of water removes most of the debris and many of the silks.

Large quantities of sweet corn are frozen. A small but increasing fraction appears on the market as corn-on-the-cob, but the greatest amount is sold as whole kernel corn. There is also a frozen counterpart of cream-style corn. Frozen succotash is packaged. Tressler *et al.* (1968) give a very thorough discussion of methods of preparing these products. In general, manufacturing techniques down to the point of packaging are rather similar for canned and frozen corn.

The use of corn as the expanded or puffed kernel is certainly very ancient, examples of grain suitable for this purpose having been found not only in the Toltec pyramids of Central America, but also in 4,000-yr old deposits in the Bat Cave of New Mexico. Popcorn has always been predominantly an American food, although fairly large quantities are sold in England, principally as caramel-coated confections, and much lesser quantities in other countries of the Eastern hemisphere.

In the United States, popcorn growing and processing are big businesses. The 209,900 acres which were harvested in 1965 yielded more than a half billion pounds of ear corn. The leading states in popcorn production are Iowa and Indiana. In addition, Ohio, Illinois, Nebraska, and Kentucky each normally plant more than 10,000 acres annually.

Cultivation procedures for popcorn are not much different from those used for other corn crops. Time of seeding, optimum soil and climate conditions, diseases, and harvesting procedures are much the same as those for dent corn. Popcorn growers usually have more trouble with weeds because their crop grows slower than dent corn and the plants are lower in stature at maturity. The new herbicides alleviate this problem considerably. The conventional 40-in. distance between hills used for dent and flint corn is also followed in planting popcorn, but the latter grain is usually seeded at a higher rate per hill, 5 kernels per hill being a common rate for check planted popcorn.

Popcorn is a favorite crop with home gardeners, and doubtless as much is produced by them as by commercial growers, although figures for the small plots do not show up in the yield records. Commercial producers of popcorn almost always operate under contract with a large merchandising concern which provides the seed and offers an assured outlet for the product. Because of this close control by relatively few persons, annual production figures are subject to great fluctuations whereas field corn production is more inflexible as a result of the greater inertia of a larger body of independent growers Popcorn sells for considerably more per bushel (70 lb of ear corn) than field corn, but the yield per acre is less and cultivation procedures are more time-consuming.

Some of the old favorite varieties of popcorn were Japanese hulless, South American, White Rice, Tom Thumb, and Yellow Pearl. Hybrid popcorns are now sown to the almost complete exclusion of the pure varieties. Most are single-cross hybrids because the plants have greater hybrid vigor and are much more uniform than the double-cross hybrids which have been so successful in field corn. Purdue and Iowa State College carry on extensive breeding programs for the development of superior hybrids, each organization growing more than 2,000 crosses annually. Hybrids are carefully graded for yield per acre, lodging tendencies, disease resistance, and amount of expansion of the kernels.

Distribution of Production

The United States produces about 40% of the world total of corn harvested for grain. US production shows signs of leveling off at about 4 billion bushels per year, but world production has been climbing steadily for many years as other countries learn advanced

TABLE 9

PRODUCTION OF CORN FOR GRAIN IN THE UNITED STATES[1]

Average 1957–59	3,409[2]
1960	3,907
1961	3,598
1962	3,606
1963	4,019
1964	3,484
1965	4,084
1966	4,117
1967 (preliminary)	4,772

[1] Anon. 1968.
[2] Millions of bushels.

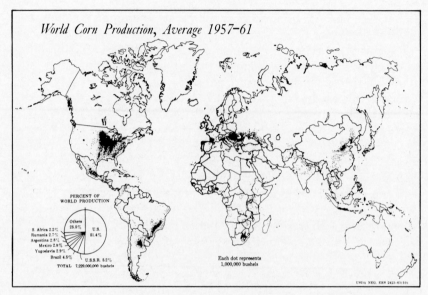

Courtesy of US Dept. Agr.

FIG. 12. WORLD CORN PRODUCTION

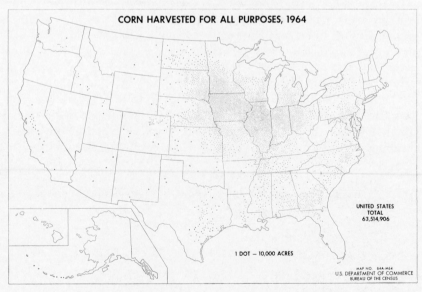

Courtesy of US Dept. Commerce

FIG. 13. CORN GROWN IN THE UNITED STATES

production methods and develop hybrid seeds. US yield per acre has been climbing rather steadily, while US and world acreages have held comparatively constant. Table 9 summarizes recent data on US production.

Corn is grown throught the United States but about $^2/_3$ is produced in an area extending westward from northern Ohio through Indiana, Illinois, and Iowa, and into adjacent portions of the contiguous states. The western boundary of intensive corn production is in the rather narrow, moist, subhumid climatic zone, while in the north it is delimited by the 70°F-average-July-temperature isotherm. This general pattern of production has remained much the same over several decades.

Iowa and Illinois vie for top honors in annual corn yields. No other state produces half as much as either of them. All states report some corn is grown, but Maine, New Hampshire, Rhode Island, Nevada, Alaska, and Hawaii produce less than a million bushels annually of corn grown for all purposes. Most of the acreage in the northernmost states is for silage.

BOTANY OF CORN

Classification

Like the other plants discussed in this volume, corn belongs in the family Gramineae. Members of this botanical group have fibrous root systems, alternate leaves, two-ranked parallel veins in the leaves, split leaf sheaths, cylindrical stems with solid nodes, and flowers in more or less chaffy spikelets.

The appropriate tribe is the Tripsaceae (Maydeae) and includes not only corn but teosinte, gamagrass, adlay, and some other less well known genera. A distinguishing characteristic of the Maydeae is the presence of male and female flowers in separate spikelets on the same plant. This tribe is similar to the sorghum group Andropogoneae from which it differs principally in the suppression of certain floral parts.

The genus *Zea* has the pistillate spikes grown together forming an ear, the grains at maturity much exceeding the glumes. According to the most prevalent taxonomic view, only one species, *mays*, is included in the genus. Some taxonomists have been of the opinion that several species, or at least subspecies, should be included in this genus. For example, Sturtevant classified sweet corn as *Zea saccharata*, and elevated other types to the level of species.

The major varieties of corn are pod corn, flint corn, dent corn, sweet corn, popcorn, flour corn, and waxy corn. All of these types

Photo by H. S. Garrison, Courtesy of US Dept. Agr.

FIG. 14.   TYPES OF CORN—REPRESENTATIVE EARS OF POP, SWEET, FLOUR,
FLINT, DENT, AND POD CORN

have the normal 2 n chromosome number of 20.   The characteristics rendering the variety distinguishable are usually endosperm differences dependent upon the operation of a single gene.   For example, the "sweet" phenotype depends upon a single recessive gene.

Pod corn is presumably a primitive type.   Each kernel is enveloped by a fibrous husk.   This character can appear in any of the other corn types described here.

Flint corn has very hard kernels, as the name indicates.   This characteristic is due to a rather thick layer of hard starch and protein just under the bran layer.   Most flints are early maturing and have a certain popularity for this reason.   The texture may adversely affect their value for feeding livestock, but presumably does not detract from their milling quality.   Flints are grown largely in Argentina and Africa.

Dent corn comprises the largest corn crop in the United States. The crown of the kernels exhibits a pronounced concavity at maturity due to shrinkage of the endosperm as moisture is lost.   The grains are hard, but not as hard as those of flint corn.

Flour corn is grown in South and Central America to a considerable extent.   The grains are large and soft, and the endosperm is very friable.   These characteristics permit easy grinding of the grain into meal, an advantage in home preparation methods.

Sweet corn and popcorn have been described in a preceding section of this chapter.

Waxy corn does not contain wax as such but owes its texture to the presence of large amounts of the amylopectin fraction of starch. It is assuming increasing importance due to the food and industrial uses which are being found for waxy starch.

Description of the Plant and Its Seed

Zea is a robust, erect annual grass.   Some varieties may reach 15 ft or more in height, while others rarely exceed $1^1/_2$ ft at maturity. The plant has broad leaves arranged in two vertical ranks.   The inflorescences are monoecious.   The flowers of the staminate tassel are borne in many spike-like racemes which together form large spreading panicles which terminate the stems.   A pistillate influorescence is borne on one or more spikes of the leaves.   There are eight or more rows of spikelets on a greatly thickened axis (cob) and the whole is inclosed in foliaceous bracts (husks).   Protruding from the tops of these bracts are the styles called silks.   The styles are long and slender and may be fertilized throughout their length. Except in pod corn, the floral bracts on the axis do not envelope the kernels.   The spikelets are unisexual.   The pistillate spikelets occur in pairs, both members of which may be fertile, or, usually, one member is fertile and one is sterile.   The glumes are broad and rounded at the apex.   Staminate spikelets are 2-flowered and are arranged in pairs on the side of a continuous rachis, 1 member of the pair being sessile and 1 pedicillate.   The glumes are membranaceous, acute, and covered with short hairs.   Hyaline palea and lemma are present (Hayes *et al*. 1934).

STRUCTURE AND COMPOSITION

Wolf *et al*. (1952A, 1952B, 1952C, and 1952D) reported an intensive investigation of the structure, macroscopic and microscopic, of the corn kernel. In their introduction, these investigators say: "The kernel of corn is a fruit composed of a thin pericarp enclosing a single seed.   The pericarp is the mature ovary wall and comprises all of the outer cell layers down to the seed coat.   Along its inner surface,

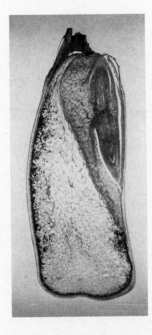

Courtesy of Dr. S. A. Watson, Corn Products Co.

FIG. 15. PHOTOMICROGRAPH OF LONGITUDINAL
SECTION OF A KERNEL OF DENT CORN

Preparation method has removed the contents of
some of the cells.

it adheres closely to the seed coat. The latter in turn encloses the
germ and the endosperm, the three forming the seed. This type of
single-seeded fruit, in which the pericarp does not open on drying to
liberate the seed, is characteristic of the cereal grains. It is known
as a caryopsis."

Figure 15 is a photomicrograph of a section of a dent corn kernel.
Figures 16 and 17 are drawings of sections with the most important
tissues labeled. The four principal parts are the tip cap, germ, hull
(pericarp), and endosperm. The color of the kernel can vary from
white to dark red or brown. The weights (dent variety) will range
from 150 to 600 mg, with an average of about 350 mg (Watson
1967).

TABLE 10

PROXIMATE ANALYSIS OF CORN GRAINS[1]

Moisture, %[2]	16.7
Starch, %	71.5
Protein (N $\times$ 6.25), %	9.91
Fat, %	4.78
Ash (oxide), %	1.42
Fiber (crude), %	2.66
Sugars, total, %	2.58
Total carotenoids, mg/kg	30.0

[1] Watson (1967). Averages obtained at 2 corn milling plants in Illinois during the period 1958–1962.
[2] Moisture is stated on a wet weight basis. All others on a dry weight basis.

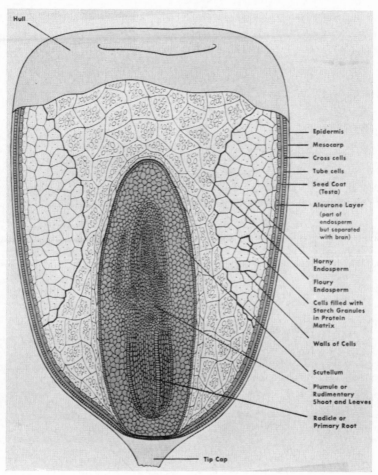

Courtesy of Wheat Flour Institute

FIG 16. DIAGRAM OF LONGITUDINAL SECTION OF A KERNEL OF CORN

Table 10 lists the average proximate analysis of corn kernels based on many samples of commercial grain received at 2 wet milling plants in Illinois during the period 1958 to 1962. These samples were predominantly hybrid dent corn. Flint and flour corn can be expected to have somewhat similar composition, but other types of corn may vary considerably in composition as shown in Table 11. These data are on an "as is," or wet, basis, and it can be seen that the sweet corn values refer to a dried kernel rather than to the "fresh" or succulent form in which the food is usually consumed. Notable differences between the varieties are apparent in the ether extract

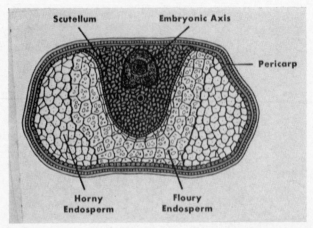

Courtesy of Wheat Flour Institute

FIG. 17. DIAGRAM OF CROSS SECTION OF A KERNEL OF
CORN

TABLE 11

COMPOSITION OF WHOLE CORN KERNELS[1]

Fraction	Sweet Corn (%)	Hybrid Popcorn (%)	Inbred Dent Corn (%)	Hybrid Yellow Dent Corn (%)
Protein	10.88	10.69	8.31	8.06
Ether extract	8.18	3.69	3.90	3.94
Crude fiber	1.99	8.25	1.74	2.09
Moisture	10.10	9.78	11.46	10.12
Ash	1.83	1.45	1.18	1.40
N-free extract	67.02	72.14	73.41	74.39

[1] Fan *et al.* (1965).

TABLE 12

DISTRIBUTION OF THE COMPONENTS OF YELLOW DENT CORN AMONG THE FRACTIONS
OF THE KERNEL[1]

	Endosperm (%)	Embryo (%)	Pericarp (%)	Tip Cap (%)
Proportion of the part to the whole kernel	82	11.6	5.5	0.8
Protein	73.1	23.9	2.2	0.8
Oil	15.0	83.2	1.2	0.6
Sugar	28.2	70.0	1.1	0.7
Starch	98.0	1.3	0.6	0.1
Ash	18.2	78.5	2.5	0.8

[1] Earle *et al.* (1946).

values, which are relatively high for sweet corn, and the crude fiber, which is much higher in popcorn than in the others.

The distribution of some of the components within the dent corn kernel is given in Table 12. There is a characteristically high level of oil in the embryo and of starch in the endosperm. There are also relatively high concentrations of sugar and ash in the embryo.

Carbohydrates and Related Compounds

Starch is the predominant component of corn. Other carbohydrates are present in relatively small amounts. In waxy varieties, close to 100% of the starch will be in the branched molecule form, i.e., amylopectin. In ordinary varieties, a substantial amount of amylose is present. The proportion of amylose averages about 27% and seems to vary within a rather narrow range. In a series of nonwaxy varieties comprising 190 samples from the United States and 167 samples from abroad, the amylose content of the starch was found to vary from 20 to 36% (Deatherage, *et al.* 1955). Varieties containing from 55 to 80% of the total starch as linear fraction have been developed.

Starch from dent corn includes rounded granules from the floury endosperm and polyhedral granules from the horny endosperm. The latter show pronounced pressure facets which are attributed to tension developed during field drying. Diameter of the granules lies between 2 and 30 microns. The hilum is centrally located, the lamellae indistinct or invisible, and the polarization crosses are of moderate intensity. High-amylose varieties tend to have grossly irregular granules which have little or no birefringence and are generally smaller than normal. Waxy cornstarch granules are similar to those of ordinary cornstarch except that the maximum diameter of the waxy granules may be a few microns larger.

The cell walls of the pericarp contain appreciable amounts of fibrous material. Cellulose and pentoglycan, in about equal proportions, are the principal constituents. The pentoglycan, when isolated, forms viscous pastes in water and may eventually have some commercial use. It contains residues of D-xylose, L-arabinose, DL-galactose, D-glucose, and glucuronic acid in a highly branched molecule.

About 0.1 to 0.3% raffinose, 0.9 to 1.9% sucrose, 0.2 to 0.5% glucose, 0.1 to 0.4% fructose, and smaller amounts of myo-inositol and glycerol are present in sound corn kernels. Maltose and probably other saccharides appear during germination while raffinose disappears (Bond and Glass 1963).

Nitrogenous Compounds

As might be expected, most of the nitrogen of the corn kernel is present in the form of protein. The proteins can be separated into various fractions based on their solubility in various reagents. The significance of such fractionations can be debated, but the predominant protein of corn, zein, is a prolamine; that is, it is soluble in dilute alcohol. The principal site of zein and other proteins is in the endosperm. The hull contains little protein Germ protein usually accounts for between 15 and 25% of the total protein of corn and may contribute 25 to 40% of the total kernel lysine.

Wolfe and Fowden (1957) obtained from East African maize purified preparations representative of the entire protein composition of the seeds. The varieties included both dent and flint forms. Under the conditions of their tests, the amino acids produced by acid hydrolysis accounted for between 96.0 and 99.6% of the total protein nitrogen. Total protein content of the original seeds (N × 6.25) varied from 8.2 to 10.3% (air-dry basis). A summary of the amino acid distibution found by these authors is given in Table 13. These figures are generally higher than those previously reported in the literature, especially for cystine, methionine, and lysine.

Relatively small amounts of nonprotein nitrogen are found in corn. The amounts of amino acids, quaternary-N compounds, nucleosides, purines, and pyrimidines in germ, endosperm, and bran

TABLE 13

AMINO ACID CONTENT OF PROTEIN ISOLATED FROM MAIZE SEEDS[1]

Amino Acid	Range[2]	Average[2]
Alanine	5.3–7.5	6.6
Arginine	8.5–14.7	11.7
Aspartic acid	3.2–5.5	4.0
Cyst(e)ine	1.8–3.0	2.4
Glutamic acid	7.5–9.5	8.5
Glycine	2.7–4.7	3.8
Histidine	4.8–7.8	6.2
Isoleucine	2.9–4.9	4.0
Lysine	4.2–7.5	5.4
Leucine	6.8–11.0	9.3
Methionine	2.6–3.3	3.0
Phenylalanine	3.5–5.6	4.3
Proline	3.3–5.0	4.5
Serine	3.2–4.1	3.6
Threonine	2.1–10.8	6.5
Tryptophan	0.72–1.05	0.83
Tyrosine	2.0–3.2	2.5
Valine	2.9–6.2	4.4
Amide	4.7–8.6	. . .

[1] Wolfe and Fowden (1957).
[2] Values are in grams of amino acid per 100 gm total protein.

fractions were measured by Christianson *et al.* (1965). It appears that over 50% of the nonprotein nitrogen in whole corn is in the form of amino acids, and they are distributed almost equally between the endosperm and the germ fractions. Only minor amounts were found in the bran. The concentration in the germ is several times that in the endosperm. The quaternary N and heterocyclic-N compounds were found primarily in the germ.

Lipids and Related Compounds

Analyses of a series of 125 inbreds showed oil content of the whole kernel to vary between 1.2 and 5.7% (Quackenbush *et al.* 1961). Varieties developed particularly for high oil content are known to yield as much as 14% of this constituent. Fatty acids which have been reported by various investigators are 56% linoleic, 30% oleic, 0.7% linolenic, stearic, palmitic, arachidic, and others in relatively insignificant amounts.

The study by Quackenbush and associates reported iodine values from 111 to 151, and tocopherols from 0.03 to 0.33%. Appreciable amounts (around 1%) of phosphatides are present. The lecithin content is on the order of $1/2$%.

The oil from the total kernel takes on many of the characteristics of the germ lipids, of course, since the embryo is the preponderant source of fatty material. Although few studies have been published on the subject, it appears that endosperm oil is similar in most respects to germ oil.

Corn oil is one of the seed oils known to have a hypocholesteremic effect in man and some animals. The property of reducing cholesterol in the blood is evidently due, at least in part, to the large amounts of plant sterols, such as β-sitosterol, in corn oil (Beveridge *et al.* 1964).

Inorganic Constituents

Cannon *et al.* (1952) summarized in Table 14 the work of several investigators on the ash components of the corn kernel. See their paper for original references. Most of the phosphorus is present as phytin, and nearly all of the phytin is in the germ.

Pigments

The system of carotenoid polyenes in yellow corn grain is quite complex, and separation of the individual compounds has proved to be difficult. Quackenbush *et al.* (1961) described a method for elution of seven fractions from a magnesia chromatogram. Spec-

TABLE 14

ASH CONSTITUENTS IN THE CORN KERNEL

Element	Found
	Ppm
Na	10–1,250
K	2,200–9,200
Cu	4–17
Ca	60–1,800
Ba	9
Mg	900–2,700
Zn	20–34
B	1
Al	130–360
Si	110–210
Ti	1.4
P	2,300–8,000
As	0.03–0.36
S	400–3,000
Se	0.1–30
F	1–2
Cl	40–4,600
Br	1.7
I	0–2.8
Mn	4–500
Fe	17–550
Co	0.01–0.10
Ni	0.14

trophotometry of these fractions provided values for 3 provitamins A as well as 8 carotenoids with little or no vitamin activity. In addition to *cis* isomers of the major polyenes, a number of minor components were observed. This method showed that wide differences exist in carotenoid and tocopherol distribution in 125 different inbred lines of corn (Quackenbush *et al.* 1961). Provitamin A ranged from a trace to 7.3 γ per gram of corn. Lutein, the predominant xanthophyll, ranged from 2 to 33 γ per gram. The total tocopherols, were between 0.03 and 0.33%. In this series the total oil contents varied from 1.2 to 5.7%, and the oil samples had iodine values of 111 to 151. There was no apparent correlation between provitamin A content and the percentage or composition of oil.

Xanthophylls of typical double-cross hybrid yellow dent corns varied from 10 to 30 ppm and carotenes from 1 to 4 ppm. Some exotic strains from South America contained about 60 ppm, the highest found to date. Analyses of several samples indicated that xanthophylls and carotenes may be independently inherited. Outward appearance of the intact seed was not correlated with xanthophyll content (Blessin *et al.* 1963).

Samples of hybrid yellow dent corn, typical of those processed by wet-millers, showed a 3- to 4-fold variation in xanthophylls (about

10–30 ppm) and carotenes (1–4 ppm). Variation in total carotenoid content (247–379 ppm) of two commercial samples of corn gluten containing 60 to 70% protein was of the same magnitude as the differences in carotenoid level (19–30 ppm) of the whole corn used for processing. The total carotenoid content of gluten feed (21% protein) varied from 14 to 34 ppm and that of gluten meal (41% protein) from 65 to 253 ppm. The gluten fractions contained larger quantities of noncarotenoid pigments than did the whole corn (Blessin *et al.* 1964).

Using a method based on chromatography and spectrophotometry which gave a sharp separation of carotenes from free xanthophylls, Blessin (1962) obtained the values reproduced in Table 15. Blessin *et al.* (1963) used a similar method to determine the xanthophylls and carotenes in hand-dissected and dry-milled fractions of yellow dent corn (see Table 16).

TABLE 15

XANTHOPHYLL ESTERS IN CORN[1]

	Nonsaponified		Saponified		
	Caro-tenes (Ppm)	Xantho-phylls (Ppm)	Caro-tenes (Ppm)	Xantho-phylls (Ppm)	Xantho-Esters (Ppm)
High-xanthophyll corn	3.0	35.9	2.6	36.3	0.4
Hybrid yellow corn	2.0	20.1	1.6	20.5	0.4
Corn gluten (60% protein)	16.9	201.6	13.2	205.3	3.7

[1] Blessin (1962).

TABLE 16

DISTRIBUTION OF CAROTENOIDS IN HAND DISSECTED YELLOW DENT CORN FRACTIONS[1]

Fraction	Variety[2]	Dry Weight (%)	Carotenes (Ppm)	Xantho-phylls (Ppm)
Whole corn	A	100.0	1.8	19.0
	B	100.0	2.0	13.9
	C	100.0	1.7	19.1
Bran	A	7.4	0.3	1.1
	B	7.6	0.3	1.8
	C	8.3	0.2	1.4
Germ	A	10.1	0.5	3.5
	B	10.0	1.0	3.5
	C	10.3	1.3	4.1
Floury endosperm	A	33.0	1.1	12.8
	B	36.1	0.3	3.0
	C	27.5	1.1	10.0
Horny endosperm	A	49.5	2.3	27.3
	B	46.3	3.2	20.8
	C	53.9	3.7	25.5

[1] Blessin *et al.* (1963).
[2] The varieties are: A = S × 20; B = Pioneer 329; C = Commercial mixture.

QUALITY

Field Corn

In discussing the factors affecting the quality of corn, it is necessary to keep in mind that they will vary depending upon the particular end use of the grain. For example, the quality factors important in sweet corn will obviously be quite different from those looked for in field corn.

Field corn is a standard item of commerce and closely defined classes and grades have been established by the Federal government. For the purposes of the Official Grain Standards of the United States, "corn" is any grain which consists of 50% or more of whole kernels of the dent or flint variety of *Zea mays*. The balance may be made up of other varieties of maize and up to 10% of other grains for which standards have been established.

Color of the grain furnishes the basis for three fundamental market classes of corn. Yellow corn includes all varieties of yellow and may not include more than 5% by weight of kernels of other colors. A slight tinge of red on kernels which are predominantly yellow shall not affect their classification as yellow corn. White corn must contain at least 98% by weight of white kernels. Grains showing slight tinges of light straw color or of pink shall be graded white. Mixed corn includes lots of corn not falling into the classes of white or yellow.

These three basic classes are further qualified as Flint, if 95% or more of the kernels (by weight) are of the flint varieties. If it meets this requirement, the grain would be classed as Yellow Flint Corn, White Flint Corn, or Mixed Flint Corn, as appropriate. "Flint and Dent Corn" is a mixture of the flint and dent varieties which includes from 5 to 95% flint. Dent corn requires no modifying adjective signifying the variety.

Each named category must also be classified into 1 of 5 numerical grades or into sample grade according to the criteria listed in Table 17.

It can be seen that quality of corn of commerce, so far as the US Standards are concerned, is based primarily on: (1) variety, (2) color, (3) test weight per bushel, (4) moisture, and (5) amount and kind of material other than kernels of corn which have not been heat-damaged.

Many of the tests required in the Standards (e.g., those for color) can only be applied by visual inspection. Their performance requires the services of qualified inspectors, and, in case of litigation based on a difference of opinion (such litigation is rather rare), the

TABLE 17

GRADE REQUIREMENTS FOR CORN[1]

Grade	Minimum Test Weight per Bushel (Lb)	Maximum Limits			
		Moisture (%)	Broken Corn and Foreign Material (%)	Total (%)	Heat-Damaged Kernels (%)
1	56	14.0	2.0	3.0	0.1
2	54	15.5	3.0	5.0	0.2
3	52	17.5	4.0	7.0	0.5
4	49	20.0	5.0	10.0	1.0
5	46	23.0	7.0	15.0	3.0
Sample[2]	...	...	...	...	...

[1] Anon. (1964).
[2] Sample grade shall be corn which does not meet the requirements for any of the grades from No. 1 to No. 5, inclusive; or which contains stones; or which is musty, or sour, or heating; or which has any commercially objectionable foreign odor; or which is otherwise of distinctly low quality.

final decision must of necessity be based upon a weighing of the evidence of two or more experts. Tests of this sort are less satisfactory from a theoretical standpoint than are more objective criteria, but they seem to function well in practice, perhaps because the producer-vendor is generally disinclined to employ an expert of his own.

When grading a lot of shelled corn, a trier or probe is used to draw a fair average sample. If it is noticed that the interior of the grain is appreciably warmer than the surroundings, the grain is said to be "heating" and the whole batch is graded "Sample Grade." Grain removed by the probe is examined carefully for insects. If insects (or their larvae) of kinds that are known to be destructive to grain are found, the word "Weevilly" is added to the grain grade. Accidental contamination with insects of other types (e.g., ants) may be ignored. The odor of the sample is noted. If it smells musty or sour, indicating extensive microbiological activity, it is graded "Sample Grade." The same rating may be assessed if it smells noticeably of insecticides, hydrocarbons, or other foreign odorants

Weight per bushel is determined by weighing a known volume (rarely a bushel) of grain and calculating the weight of 2,150.42 cu in.

A sieve with round holes $^{12}/_{64}$ in. in diameter is used to remove cracked corn and foreign material. Any material other than corn remaining on top of the sieve is removed and added to the screenings. The material removed and that passing through the screen is weighed and the percentage by weight is calculated to determine the cracked corn and foreign material.

The kernels which have been damaged mechanically or by heating are picked from a 250-gm portion of the cleaned corn and are weighed and the percentage calculated. Heat damaged kernels are segregated from the total damaged portion and weighed separatedly. Heat damage is manifested by a darkening of the germ and other changes. It destroys the viability of the grain and reduces its milling quality.

Only the test required for fixing the lowest possible grade need be made. Thus, corn which is sour or musty immediately falls into Sample Grade and the tests for damaged kernels, bushel weight, foreign material, etc., need not be considered.

A water oven method of determining moisture percentage is the legally prescribed technique, but in practice, moisture may be determined in several ways. The AOAC Methods (Anon. 1965) require heating a sample at 266 ± 5° F and atmospheric pressure for 1 hr, or at 208 to 212°F and 25 mm pressure until constant weight is attained (about 5 hr).

Other "official" tests not required by the Standards, but which are occasionally run for special purposes are ash, extract soluble in cold water, crude fiber, crude fat, fat acidity, and total protein. Details of these tests may be obtained by referring to the Official and Tentative Methods of the Association of Official Agricultural Chemists (Anon. 1965).

The quality of field corn intended for specific purposes must take into consideration factors not listed in the Grain Standards. The percentage of protein is an important consideration in selecting corn for mixed animal feeds. The amount and kind of starch is of interest to the wet-milling industry. Freedom from any off-flavors is essential for the breakfast cereal and lye hominy manufacturers. Gelatinization or baking characteristics are often determined by the miller of corn meal and grits for human consumption.

Sweet Corn

The two principal factors contributing to sweet corn acceptability are apparently sweetness and tenderness. Tests to measure these factors must be rapid and accurate, since maturation proceeds very rapidly in many varieties of sweet corn and relatively small changes, percentagewise, can cause appreciable changes in the sensory quality of the product. For example, Geise (1953) determined the smallest increment in certain quality test results which could be detected subjectively by experienced judges at a 95% level of confidence. It was found that these increments were 0.52% for moisture, 0.06% for pericarp, 0.95% for alcohol insoluble solids, and 0.68 ml for differences in volume of 20 kernels. Most practical tests for determining the suitability of the raw material for processing are based on measurements of the hardness of the kernels, or of the moisture content. These tests range from very simple ones, as in the common thumbnail test used by experienced field men, to relatively well instrumented ones as in the succulometer test or in moisture tests by electronic devices.

The thumbnail test involves pressing the kernels while they are still on the cob and judging the "juiciness" and softness of the corn. Another visual and subjective test is based on the number of dented kernels on the ears. Ears showing kernels with concavities in the crown, indicating loss of moisture, are too far advanced in maturity to process properly into a high quality product. According to Huelsen (1954), lots of Country Gentleman and Evergreen varieties in which more than 5% of the ears have one or more dented kernels are not suitable for packing as fancy cream style corn. Corn at this stage contains about 60% moisture.

Henry *et al.* (1956) showed that the desirability of the flavor, skin texture, endosperm texture, and color decreased as the corn matured to less than 76% moisture. When the moisture fell below 70%, these adverse changes were accelerated. The percent reducing sugars and total sugars traced a downward course as the corn matured.

Caldwell (1939) suggested the use of a puncture test to determine toughness of the pericarp since it is known that the pericarp becomes thinner and harder as the corn matures. However, there are differences in pericarp toughness between varieties at any given stage of maturity (Haber 1931).

Henry *et al.* (1956) surveyed a number of the methods which have been suggested for use as indicators of sweet corn maturity. Results of the various moisture tests studied—Steinlite moisture meter, succulometer, percentage of moisture by the vacuum oven method or calcium carbide method, and the per cent soluble solids by refractometer were well correlated with the degree of maturity. Measurement of the turbidity of the juice was found to be a rather impractical test for control laboratory use although the results of the test appeared to be correlated with maturity down to about 70% moisture content. The ethanol-insoluble solids content was a good indicator of the degree of maturity and the values for raw and canned samples agreed rather well.

Table 18 lists the correlation of various objective tests applied to sweet corn with the subjective preference ratings.

Table 18

CORRELATION COEFFICIENTS BETWEEN FLAVOR AND APPEARANCE PREFERENCE
SCORES AND THE MEASURED QUALITY FACTORS OF SWEET CORN[1]

	Canned Corn	
Factor	Range of Means	Corr. Coef. Flavor
Preference panel		
Flavor	3.9–7.4	. . .
Trained panel		
Sweetness of product	2.0–7.5	0.68[2]
Flavor intensity	1.9–7.4	0.73[2]
Tenderness	2.0–8.6	0.66[2]
Maturity	2.0–7.1	0.68[2]
Physicochemical tests		
Hunter Rd	34.4–43.5	0.31
a	−1.0–4.8	−0.30
b	28.4–34.0	−0.14
Shear resistance	0.91–1.65	−0.46[2]
Soluble solids, liquid	8.5–12.5	−0.07
% moisture, kernels	70.1–80.7	0.51[2]
Soluble solids, kernels	19.0–30.0	−0.51[2]
USDA grade		
Total score	82–95	0.40[2]
Maturity	31–39	0.46[2]
Color	8–10	−0.05
Cut, uniformity	8–10	−0.14
Absence of defects	16–19	0.22
Flavor	17–19	0.34

[1] Sather and Calvin (1963).
[2] Significantly different from zero at $p = 0.01$.

None of the usual tests for sweet corn maturity are sufficiently rapid and accurate to satisfy commercial users. Wei *et al.* (1967) developed 2 "improved" tests, 1 based on density of expressed juice by pycnometer and the other on refractive index of starch-free juice. Results were comparable to those obtained with other objective tests and with organoleptic evaluations.

The density of the juice was highly correlated with vacuum oven moisture but the former value could be obtained much more rapidly (15 min vs. 24 to 96 hr). Density changed 0.01 units for each 2% in moisture content, pointing to a potential accuracy of 0.02% moisture. Their data indicate that density measurement will vary within ±1.33% of the true moisture content in 95 % of the cases.

Results with the refractometer method were not as precise but showed reasonable agreement (correlation of about 0.79) with vacuum moisture determination.

Organoleptic evaluations of the boiled corn kernels were made, using a 9-point scale, and compared with the objective tests run by these investigators. Results are shown in Table 19. The best correlations with the taste test values were shown by the density test and the vacuum oven moisture determination.

TABLE 19

CORRELATION BETWEEN OBJECTIVE AND ORGANOLEPTIC DETERMINATIONS OF
SWEET CORN FROM ABOUT EIGHT PICKINGS FROM EACH OF TWO PLANTINGS OF
TWO VARIETIES DURING ONE YEAR[1]

Objective Determination, 1	Organoleptic Evaluation, 2	Correlation Coefficient, r	% Variation in 2 Accounted for by 1, r^2
Vac. oven moist.	Flavor	0.861[2]	74.1
Density	Flavor	−0.944[2]	89.1
Sugar	Flavor	0.760[2]	57.8
Maturity index	Flavor	−0.845[2]	71.4
Vac. oven moist.	Texture	0.853[2]	72.8
Density	Texture	−0.753[2]	56.6
Sugar	Texture	0.643[2]	41.3
Maturity index	Texture	−0.756[2]	57.1
Vac. oven moist.	Color	0.595[2]	35.4
Density	Color	−0.552[2]	30.5
Sugar	Color	0.351	12.3
Maturity index	Color	−0.498[3]	24.9

[1] Wei et al. (1967).
[2] Significant at 1% level 0.526.
[3] Significant at 5% level 0.413.

Popcorn

Although commercial distributors of popcorn claim that they consider the kernel texture, flavor, color, and other subjective characteristics in selecting seed varieties, the chief basis for judging popcorn desirability has always been the relative amount of kernel expansion that is obtained when the corn is popped. This is a natural tendency, since expansion is closely related to the price received by the processor. Popped corn is usually sold and bought by the consumer (in amusement parks, theaters, etc.) on the basis of volume rather than weight. The concessionaire must fill a box or bag of given volume with popped corn, regardless of whether he has a raw popcorn of high or low expansion. Naturally, he prefers a corn of high expansion since this means he will have to purchase fewer pounds of raw material to fill the required number of containers. Furthermore, corn of higher expansion potential generally has a more tender texture when popped than does corn of lower expansion.

The trade measures expansion by popping a given weight or volume of corn under controlled conditions, dumping the popped kernels into a graduated plastic cylinder of standard diameter and reading the height of the column. Formerly, a standard volume of kernels was used for test popping, but measurement in this manner did not always accurately express the contribution of kernel weight to the results, so a change was made in the procedure, and, currently, a standard weight of grain is used in the test. Results are now expressed as cubic inches of popped corn obtained per pound of raw

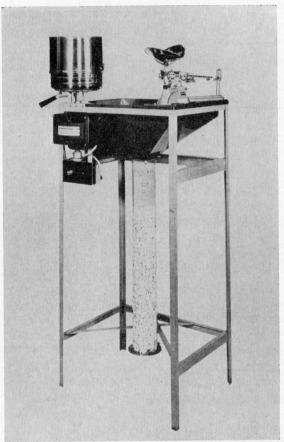

Courtesy of C. Cretors Co.

FIG. 18. OFFICIAL WEIGHT-VOLUME TESTER FOR POPCORN

corn. In the volume versus volume test the volume of unpopped corn used filled the graduated tube to a height of 1 in., and the height of the column of popped corn in inches was the expansion ratio of the corn. Most commercial corn gives an expansion between 30- and 35-fold, although a few recently developed hybrids give expansion ratios as high as 40-fold (Nelson 1955).

It has been suggested (Richardson 1957) that pericarp thickness should be used as another indicator of popcorn quality. In this test, the pericarp sections are mounted on edge in modeling clay and observed under a compound microscope equipped with an ocular micrometer. Hull sections are more easily obtained from popped samples than from unpopped kernels. The pericarp thickness is expressed in microns, and in a large series of hybrid corn samples varied between 41 and 65. The measurement is thought to give an indication of quality because it is related to the amount of tough, horny material remaining on the popped corn and thus is related to the acceptability of the product. Pericarp thickness is largely influenced by genetic factors although changes in the environment have some effect (Richardson 1958).

The optimum moisture percentage for popping has been recommended as 13.5%. In working with 4 hybrids, Huelsen and Bemis (1954) found that increasing the moisture, at least to about 14.0%, increased expansion. However, the response to increasing moisture differed with respect to maturity at harvest and popping temperatures. Corn harvested at greater maturity (e.g., 15.65% moisture) gave higher popping expansion than a lot harvested at 31% moisture. The temperature of the popper giving optimum expansion varied with the different hybrids. Reducing the moisture content at popping likewise reduced the required popping time. There was an inverse relationship between popping expansion and time required to pop. The temperature at which popping began decreased with increasing moisture content.

FACTORS AFFECTING YIELD AND QUALITY

Soils and Climate

Corn can be grown in a wide variety of climates and on very diverse kinds of soils as indicated by the geographical distribution of its production. However, it requires deep dark silt loam, good drainage, abundant moisture, and moderately high temperatures for maximum yields.

According to Shrader and Pierre (1967), the ideal soil should have a pH of about 6.5, an exchange capacity of about 20 me per 100 gms soil, a percentage base saturation of 75 to 90% (with less than 10% of the bases being sodium), a water-holding capacity of about 2 in. of water in available form per foot of soil, and a bulk density of not more than 1.3.

Sandy soils are less suitable for production of corn because they do not retain moisture for a sufficient length of time. In addition, sandy soils often contain low concentrations of essential nutrients. The latter condition can, of course, be corrected temporarily by fertilization, but excessive leaching of the fertilizer by rainfall may make this practice uneconomical. Clay soils also give relatively low yields because they are liable to be too compact to allow maximum root proliferation and may have inadequate drainage.

Bates (1955) showed that the mean maximum temperature, the mean relative humidity, and the evaporation in June (the month in which corn is usually pollinated in the region where Bates conducted his experiments) were very closely correlated with corn yields. Each of these three factors was more closely correlated with corn yields then was rainfall of any period of the year. The number of rains in June showed a higher correlation with yield than did any other rainfall factor. Under normal conditions in the corn belt, moisture requirements of corn plants exceed the precipitation during the peak growth period.

The rather indirect relationship of rainfall in any short period to yield may be partially explained by data of Russell and Danielson (1956), who showed that corn utilized water to a depth of 5 ft or more in a deep Brunizem soil in Illinois, while rainfall and irrigation affected the soil moisture profile to a depth of only 2 ft on both corn and fallow plots. Letey and Peters (1957) showed that corn yields were closely related to the reserve soil moisture conditions at the beginning of the growing season and to the soil moisture stress to which the plant was subjected during the growing season. The moisture stress effect on growth was found to be closely related to the seasonal weather. The efficiency of water use was found to be strongly conditioned by soil temperature.

Gingrich and Russell (1956) found that both the soil moisture tension and the oxygen concentration of the root atmosphere had important effects on growth.

Willis *et al.* (1957) grew corn seedlings in temperature controlled greenhouses and in field plots where the temperature was adjusted by underground heating cables. In the average soil temperature range tested, 60° to 80°F, the rate of corn growth seemed to be approximately in accordance with the Van't Hoff law, the Q_{10} of the law being from 2.0 to 2.8. The most favorable soil temperature at the 4-in. depth for corn grown in central Iowa appeared to be about 75°F. The time between silking and maturity was reduced by increased soil temperature.

Corn is grown in many areas where the temperature and soil requirements for optimum yield are not found. Use of hybrids selected to fit the limitations of the soil and the climate where optimum conditions do not exist tends to reduce, but does not eliminate, the decrease in yield resulting from these limitations. Proper selection of planting time, to take advantage of a soil temperature of at least 60°F is also helpful, but this procedure has obvious limitations.

Varieties

Hybrids, which constitute the majority of corn seed now used in the United States, vary considerably in their growth rate, resistance to disease and insects, ratio of stover to ear, etc. It is evident, therefore, that the hybrid selected for seed will have a considerable effect on the yield per acre. Duncan (1967) points out that appropriate hybrids are those which allow weather, technology, and management to be fully expressed in high yields.

Information on yield and lodging resistance of hybrids can often be obtained from the corn yield tests conducted in many states. It is also desirable to know about insect and disease susceptibility, suitability for combining, maturity rating, and dry-down characteristics. Such data can sometimes be obtained from representatives of the companies supplying the seed.

Rotation and Fertilizer

Corn extracts enormous quantities of nutrients from the soil. In fact, it is the cereal crop most destructive to soil fertility. Millar and Turk (1951) indicate that a 50-bu per acre corn crop removed from each acre of soil 78.4 lb of nitrogen, 27.6 lb of phosphorus, 55.2 lb of potassium, 14.7 lb of calcium, and 5.6 lb of magnesium. For this reason, heavy fertilization is essential if corn is to be grown several years in succession on the same acreage. On the Morrow field plots of the University of Illinois (De Turk *et al.* 1927), where corn has been grown continuously since 1888, the yield has shown a steady (though uneven) decrease through the years, falling from an initial value of about 54 bu per acre to about 24 bu per acre. When oats were grown in alternate years, corn production held up slightly better, showing a decrease from about 50 to about 28 bu per acre. A corn-oats-clover rotation gave much better results, with only a slight decline throughout the test period. The corn-oats-clover pattern, together with applications of manure and phosphated lime actually showed a significant increase in yield during the earlier years and subsequent maintenance of production at a high level.

Chen and Arny (1941) studied yields of corn grown continuously for 30 yr on plots at St. Paul, Minn. The figures are somewhat erratic, but show that yields near the end of the study were not much different from the earliest figures when the plots were manured at the rate of two tons per acre per year.

Richer (1950) analyzed results of fertilizer applications on plots where a corn, oats, wheat, and hay rotation had been maintained since 1881.   Some of his conclusions were: (1) Lime alone does not increase yields; (2) High yields can be maintained indefinitely with inorganic commercial fertilizers; (3) Manure should be reinforced with superphosphate; and (4) Fertilizers alone cannot compensate for the deficiency of an acid soil, but the addition of lime can increase the efficiency of fertilizers on these soils up to 300% as measured by yield.

Lang *et al.* (1956) showed that the protein content of the kernels of corn increased with increasing nitrogen supply in the soil.   The oil content also appeared to increase slightly with added fertilizer. Prince (1954) showed that, under the conditions used by him, there was a direct relationship between the amount of nitrogen applied to the soil and the contents of crude protein, zein, and leucine in the grain.   Increasing the rate of nitrogen resulted in an increased percentage of leucine in the grain and in the crude protein.   On the other hand, increasing the plant stand decreased the total leucine content.   This decrease in leucine content was not proportionately as great as the decrease in crude protein.   Therefore, the net effect was to increase the percentage of leucine in the protein.   There was not much effect on tryptophan content of nitrogen application or of population.   Prince concluded: "The variation in content of the amino acids discussed above suggests that nitrogen fertilization in relation to plant population, as well as variety, has an important effect on protein composition."   On the other hand, Zuber *et al.* (1954) found that application of 50 lb of nitrogen under certain conditions actually resulted in a lowering of the protein content of the grain although protein content of the stover increased.   Evidently, conditions must be properly chosen if grain of higher protein content is to result from applications of nitrogenous fertilizers.

Population

It might be expected that grain yield would reach a peak and then decline as the plant population per unit area was increased.   This effect does exist, at least under some conditions.   Haynes and Sayre (1956) planted corn in rows $8^1/_2$ ft apart to reduce row competition to a low level, and then varied the within-row spacings of plants. There was little difference between total plant weights per acre as the within-row spacing was varied from 1 to 4 in., but the closer spacings increased the stover to ear ratio.   Ear corn yield per acre was greatest at a plant spacing of about 4 in.   Genter *et al.* (1956) showed that planting at 16,000 per acre resulted in higher protein content of the grain than did planting at 10,000 plants per acre.   Oil content of the kernels was virtually the same at both levels of population.

Duncan (1958) found that the log of the average yield of individual corn plants making up a population bears a linear relationship to the population, at least between about 5,000 and 25,000 plants per acre. Therefore, only two yield population values are needed to estimate yields at any other population within the linear range.

Where phosphorus and potassium were not limiting, Thomas (1956) found that the average weight of the ears of corn decreased as the planting rate was increased, and this trend could not be reversed by the application of nitrogenous fertilizers.

Cultivation

Tillage practices intended to optimize conditions for germination and to prevent competition by weeds have undergone some changes in recent years. These changes, mainly aimed at reducing the amount of labor required to secure maximum yields, have been made possible by the availability of new kinds of equipment and by the development of selective herbicides.

Weeds compete with the crop for moisture, nutrients, and solar energy. Vengris et al. (1955) found that weeds competed strongly with corn at all rates of fertilizer (N, P, or K) application, and suppressed growth and yield of corn. They stated: "The feasibility of maintaining high corn yields in the presence of competing weeds by increasing the rate of fertilization is strongly questioned." Certain weed species were much more adversely affected by inadequate essential nutrients than was corn. Thus, in some instances, corn with weeds yielded better when no fertilizer was applied. Staniforth (1957) found that this did not apply to mature yellow foxtail (Setaria lutescens (Weigl)) infestations. Applications of 0, 70, and 140 lb of nitrogen per acre resulted in reductions of 14, 10, and 5 bu per acre from the yield in uninfested plots. Corn yields were increased 2 to 3 times more than the foxtail by fertilizer applications.

Weed growth is conventionally controlled by turning over or breaking up the soil with disk-harrows or field cultivators. There are definite possibilities of eliminating the usual primary tillage and post-planting tillage in certain types of soil when weed growth is reduced by the use of selective herbicides applied at carefully chosen times. Such techniques are by no means universally applicable, however, and require an initially good soil condition.

According to Larson and Blake (1967), hybrids should be planted as early as soil conditions and climate permit. In Michigan (East Lansing), for example, this would be the first week in May, generally. In the central corn belt, the last week in April is a suitable planting time. Later plantings tend to give lower yields, have higher moisture content at harvest, and suffer more insect damage. The improvement due to early planting is usually attributed to improvement in available soil moisture at critical growth periods.

Insects and Diseases

Insects which are primarily destructive to the seed are seed corn maggots (*Hylema platura* Meigen), seed corn beetles (*Agonoderus lecontei* Chaudoir), and *Clivina impressifrons* Le Conte), thief ants (*Solenopsis molesta* Say), and wireworms (*Elateridae* spp.). Chemical controls are usually effective. Timing of planting to insure rapid germination will minimize loss of seed due to these pests.

Damage to corn roots is usually the result of the activities of corn rootworms (*Diabrotica* spp.), grape colapsis (*Colapsis flavida* Say), Japanese beetle (*Popillia japonica* Newman), white grubs (*Phyllophaga* spp.) annual white grubs (*Cyclocephala* spp.), and corn root aphid (*Anuraphis maidiradicis* Forbes). Most of these insects, except annual white grubs, can be partially reduced by crop rotation, but use of insecticides is generally the method of choice for control.

Among the insects which feed on the underground portion of the stalk are the sod webworm (*Crambus* spp.), the black cutworm (*Agrotis ipsilon* Hufnagel), and billbugs (*Sphenophorus* spp.). Crop rotation and improvement in drainage are cultural methods which have been suggested for control.

Exposed corn leaves are attacked by many insects, such as corn flea beetles (*Chaetocnema pulicara* Melsheimer), yellow and black grass thrips (*Anaphothrips obscuras* and *Frankliniella tenuicornis*), corn blotch leaf miners (*Agromyza parvicornis* Loew), yellow-striped armyworm (*Prodenia ornithogalli* Guenee), two-spotted spider mite (*Tetranychus urticae* Koch), true armyworms (*Pseudaletia unipuncta* Haworth), caterpillars (*Simyra henrici* Grt.), grasshoppers (*Melanopus* spp.), chinch bugs (*Blissus leucopterus* Say), corn leaf beetles (*Oulema melanopus* Linnaeus), and corn leaf aphids (*Rhopalosiphum maidis* Fitch). Control is generally by insecticides.

Insects feeding on the whorl, stalk, or ear include European corn borers (*Ostrinia nubilalis* Hübner), the southwestern corn borers (*Zeadiatraea grandiosella* Dyar), southern corn stalk borers (*Diatraea crambidoides* Grote), common stalk borers (*Papaipema nebris* Guenee), fall armyworms (*Spodoptera frugiperda* J. E. Smith), corn earworms (*Heliothis zea* Boddie), corn sap beetles (*Carpophilus dimidatus* Fabricius), and scavenger beetles (*Glischrochilus quadrisignatus*).

For further details on insect pests of corn, the specific damage attributable to them, and methods of control, the reader may consult the very thorough report of Petty and Apple (1967), on which the preceding discussion is based.

Corn seems to be relatively resistant to disease, as compared to other grain crops. The yearly loss from this cause in the United States has been estimated at from 2 to 7%. In some cases, localized losses due to leaf diseases may be as high as 60%.

Corn is subject to parasitic diseases caused by fungi, bacteria, and viruses, and to nonparasitic diseases such as those caused by nematodes and by nutrient element deficiencies. Seed rots and seedling blights can be caused by *Pythium* spp., *Diplodia maydis*, *Gibberella zeae*, *Penicillium oxalicum*, and *Rhizoctonia bataticola*. Stalk rots can also be caused by *D. maydis* and *G. zeae*. Charcoal rot of stalks and roots is due to *R. bataticola*.

The two smut diseases infecting corn are known as head smut and common (or boil) smut. The latter is due to *Ustilago maydis*, while head smut is caused by *Sphacelotheca reiliana*. Damage can be reduced by using resistant hybrids, although there are none which are immune.

Leaf diseases are caused by a number of fungi and bacteria, but the only diseases of this type which are of economic importance in the United States are northern corn leaf blight (*Helminthosporium turcicum*), southern corn leaf blight (*Helminthosporium maydis*), and bacterial wilt (*Xanthomonas stewartii*). When possible, control is best achieved through planting resistant hybrids.

Corn rust and southern rust occur, but do not do much damage in this country. Brown spot, caused by the fungus *Physoderma maydis*, is found occasionally, mostly in the southeastern states.

Other diseases of economic importance include ear rot, caused by several genera of fungi, crazy top (*Sclerophthora macrospora*), and virus diseases sugarcane mosaic, celery stripe, leaf fleck, and stunt. For further details, see Ullstrup (1967).

NUTRITIONAL ASPECTS

In some "underdeveloped" areas, corn constitutes a substantial part of the diet of most of the population. This situation was at one time prevalent throughout much of the southern United States but economic improvement and education on dietary factors have reduced the problem. In South and Central America, corn is still the staple food. Physiological disorders attributable to nutritional deficiencies can be observed in such populations. As a result, considerable effort has been devoted to the study of the nutritional adequacy of corn. It has been well documented that the limiting factor in the growth promoting ability of corn protein is the relatively low content of lysine residues, while niacin is the limiting vitamin.

In Central and South America, corn is generally consumed in the form of tortillas, i.e., flat thin disks of baked masa. Preparation of masa involves heating and soaking the whole kernels in lime water, washing the cooked kernels to remove most of the lime, and grinding the softened endosperm to a paste or dough.

In a typical procedure, 1 unit volume of corn kernels will be added to 2 volumes of a lime solution. The solution will contain, on the average, about 1% of lime. The mixture of lime water and corn is heated to about 180°F for 20 to 45 min, and then allowed to cool while standing overnight. The fluid is decanted the next morning, and the corn is washed 2 or 3 times with water. The germ is generally not removed in this step. The corn is ground to a fine dough called "masa" which is shaped into thin patties of about 50 gm weight and cooked on both sides on a hot iron plate (Bressani *et al.* 1958).

It has been shown (Bressani *et al.* 1958) that substantial percentages of nutrients are lost during the preparation of masa. For white corn, the combined physical and chemical losses averaged 60% of the thiamine, 52% of riboflavin, 32% of niacin, 44% of the ether-extractables and 10% of nitrogenous substances. Comparative figures for yellow corn are 65%, 32%, 31%, 33%, and 10%. Yellow corn also loses 21% of the carotene originally present. The lime-heat treatment does, however, increase the rate of release of most of the essential amino acids (Bressani and Scrimshaw 1958).

Several workers (for example, Harper *et al.* 1958) have confirmed that alkali-treated corn, when fed to rats *ab libitum* will prevent the development of the symptoms of niacin deficiency which are associated with the consumption of untreated grain. The beneficial effect of alkali treatment is evidently due to the release of niacin from a protected, bound, or otherwise unavailable form. The "potential" niacin can also be released by prolonged boiling.

Results reported by Truswell and Brock (1961) on the effect of supplementing corn protein with purified amino acids indicated that lysine is the first limiting amino acid. According to Kies *et al.* (1965), the results might have been more informative if the total nitrogen intake had been maintained at an optimal level or if the effect of nonessential amino acid intake on the nitrogen retention had been followed. These workers showed that the nitrogen retention of adult men fed isocaloric diets was significantly greater when white corn meal provided 8.0 gm nitrogen per subject per day than when lower amounts were ingested. When a suboptimal intake (6.0 gm N/subject/day) was fed, the optimal N-retention level could be reestablished by the addition of 2.0 gm nitrogen from any of several purified essential amino acids or even certain other purified sources of nitrogen. Therefore, it would appear that nitrogen from any metabolically usable source is the first limiting nitrogeneous factor in corn protein, as determined by nitrogen retention in adult men.

TABLE 20

AVERAGE VITAMIN CONTENT OF CORN[1]

	Yellow Corn Mg/Lb	White Corn Mg/Lb
Carotene	2.20	...
Vitamin A	1990	...
Thiamine	2.06	2.22
Riboflavin	0.60	0.61
Niacin	6.40	6.04
Pantothenic acid	3.36	...
Vitamin E	11.21	13.93

[1] Ellis and Madsen (1943).

De Muelenaere *et al.* (1967A) investigated factors influencing estimation of the availability to rats of threonine, isoleucine, and valine in corn. Availability of threonine was influenced by changes in composition of the basal diet and by the amounts of protein and calories in it. The interaction between the calorie and protein levels as indicated by growth rate was eliminated by calculating availability values from a standard curve based on the amount of amino acids consumed rather than the amino acid content of the diet. Low utilization of zein and corn gluten, when assessed by the growth method was found to be mainly due to antagonism between leucine and isoleucine in the experimental diet. Elimination of this antagonism resulted in higher availability values. Although a considerable amount of valine was excreted in the feces of rats fed zein, the values for availability assessed by the growth method increased when the leucine content of the experimental diet was decreased.

In further work De Muelenaere *et al.* (1967B) studied factors affecting availability of lysine in corn and rice, as determined by growth and fecal analysis methods. In general, the lysine of proteins from these two grains was found to be highly available. Values obtained by the growth method were influenced by changes in the composition of the diet and the method of calculating availability. The values were most reproducible and least influenced by other factors when availability was calculated as a function of lysine consumption rather than a function of the lysine level in the diet. The fecal analysis method gave somewhat lower values for corn products.

BIBLIOGRAPHY

ANON. 1964. Official Grain Standards of the United States. US Dept. Agr. SRA-AMS—*177*.

ANON. 1965. Official and Tentative Methods. Association of Official Agricultural Chemists, Philadelphia.

ANON. 1968. National Food Situation. US Dept. Agr. NFS-*124*.

BATES, R. P. 1955. Climatic factors and corn yields in Texas blacklands. Agron. J. *47*, 367–369.

BEVERIDGE, J. M. R., HAUST, H. L., and CONNELL, W. F. 1964. Magnitude of the hypocholesterolemic effect of dietary sitosterol in man. J. Nutr. *83*, 119–122.

BLESSIN, C. W. 1962. Carotenoids of corn and sorghum. I. Analytical procedure. Cereal Chem. *39*, 236–242.

BLESSIN, C. W., BRECHER, J. D., and DIMLER, R. J. 1963. Carotenoids of corn and sorghum. X. Distribution of xanthophylls and carotenes in hand-dissected and dry-milled fractions of yellow dent corn. Cereal Chem. *40*, 582–586.

BLESSIN, C. W., BRECHER, J. D., and DIMLER, R. J. 1964. Carotenoids of corn and sorghum. VI. Determination of xanthophylls and carotenes in corn gluten fractions. Cereal Chem. *41*, 543–548.

BLESSIN, C. W. *et al.* 1963. Carotenoids of corn and sorghum. III. Variation in xanthophylls and carotenes in hybrid, inbred, and exotic corn lines. Cereal Chem. *40*, 436–442.

BOND, A. B., and GLASS, R. L. 1963. The sugars of germinating corn (*Zea mays*). Cereal Chem. *40*, 459–466.

BRESSANI, R., PAZ Y PAZ, R., and SCRIMSHAW, N. S. 1958. Chemical changes in corn during preparation of tortillas. J. Agr. Food Chem. *6*, 770–774.

BRESSANI, R., and SCRIMSHAW, N. S. 1958. Effect of lime treatment on *in vitro* availability of essential amino acids and solubility of protein fractions in corn. J. Agr. Food Chem. *6*, 774–778.

CALDWELL, J. C. 1939. Factors influencing quality of sweet corn. Canning Trade *22*, No. 5, 7–8.

CANNON, J. A., MACMASTERS, M. M., WOLF, M. J., and RIST, C. E. 1952. Chemical composition of the mature corn kernel. Trans. Am. Assoc. Cereal Chemists *10*, 74–97.

CHEN, H. Y., and ARNY, A. C. 1941. Crop rotation studies. Minn. Agr. Expt. Sta. Tech. Bull. *149*.

CHRISTIANSON, D. D., WALL, J. S., and CAVINS, J. F. 1965. Location of nonprotein nitrogenous substances in corn grain. J. Agr. Food Chem. *13*, 272–276.

DEATHERAGE, W. L., MACMASTERS, M. M., and RIST, C. E. 1955. A partial survey of amylose content in starch from domestic and foreign varieties of corn, wheat, and sorghum and from other starch-bearing plants. Trans. Am. Assoc. Cereal Chemists *13*, 31–42.

DE MUELENAERE, H. J. H., CHEN, M. L., and HARPER, A. E. 1967A. Assessment of factors influencing estimation of availability of threonine, isoleucine, and valine in cereal products. J. Agr. Food Chem. *15*, 318–323.

DE MUELENAERE, H. J. H., CHEN, M. L., and HARPER, A. E. 1967B.
Assessment of factors influencing estimation of lysine availability in
cereal products. J. Agr. Food Chem. *15*, 310–317.
DE TURK, E. E., BAUER, F. C., and SMITH, L. H. 1927. Lessons from
the Morrow plots. Ill. Agr. Expt. Sta. Bull. *300*.
DUNCAN, E. R. 1967. Problems relating to selection of hybrid seed;
calendarization a consideration. *In* Advances in Corn Production,
W. H. Pierre (Editor). Iowa State Univ. Press, Ames, Iowa.
DUNCAN, W. G. 1958. The relationship between corn population and
yield. Agron. J. *50*, 82–84.
EARLE, F. R., CURTIS, J. J., and HUBBARD, J. E. 1946. Composition of
the component parts of the corn kernel. Cereal Chem. *23*, 504–511.
ELLIS, N. R., and MADSEN, L. L. 1943. The vitamin content of animal
feedstuffs. US Dept. Agr. Bur. Animal Ind. AHD Publ. *61*.
ERWIN, A. T. 1951. Sweet corn—mutant or historic species. Econ.
Botany *5*, 302–306.
FAN, L., CHEN, H., SHELLENBERGER, J. A., and CHUNG, D. S. 1965.
Comparison of the rates of absorption of water by corn kernels with and
without dissolved sulfur dioxide. Cereal Chem. *42*, 385–396.
GEISE, C. E. 1953. Influence of objective quality factors on subjective
evaluation of canned sweet corn. Food Technol. *7*, 15–20.
GENTER, C. F., EHEART, J. F., and LINKOUS, W. N. 1956. Effects of
location, hybrid fertilizer, and rate of planting on the oil and protein
contents of corn grain. Agron. J. *48*, 63–67.
GINGRICH, J. R., and RUSSELL, M. B. 1956. Effect of soil moisture ten-
sion and oxygen concentration on the growth of corn roots. Agron. J.
48, 517–520.
HABER, E. S. 1931. Structure of sweet corn kernel as an index of quality.
Canner *72*, 8.
HARPER, A. E., PUNEKAR, B. D., and ELVEHJEM, C. A. 1958. Effect of
alkali treatment on the availability of niacin and amino acids in maize.
J. Nutr. *66*, 163–172.
HAYES, H. K., ENZIE, W. D., and VAN ESELTINE, G. P. 1934. The
Vegetables of New York. N. Y. Agr. Expt. Sta. Bull. *311*
HAYNES, J. L., and SAYRE, J. D. 1956. Response of corn to within-row
competition. Agron. J. *48*, 362–364.
HENRY, C. H. *et al.* 1956. Evaluation of certain methods to determine
maturity in relation to yield and quality of yellow sweet corn grown
for processing. Food Technol. *10*, 374–380.
HUELSEN, W. A. 1954. Sweet Corn. Interscience Publishers, New
York.
HUELSEN, W. A., and BEMIS, W. P., 1954. Temperature of popper in
relation to volumetric expansion of popcorn. Food Technol. *8*, 394–399.
HUNT, T. F. 1915. The Cereals in America. Orange Judd Co., New York.
KIES, C., WILLIAMS, E., and FOX, H. M. 1965. Determination of the
first limiting nitrogenous factor in corn protein for nitrogen retention
in human adults. J. Nutr. *86*, 350–356.
LANG, A. L., PENDLETON, J. W., and DUNGAN, G. H. 1956. Influence
of population and nitrogen levels on yield and protein and oil contents
of nine corn hybrids. Agron. J. *48*, 284–289.

LARSON, W. E., and BLAKE, G. R. 1967. Seedbed and tillage requirements. *In* Advances in Corn Production. Iowa State Univ. Press,
Ames, Iowa.

LETEY, J., and PETERS, D. B. 1957. Influence of soil moisture levels and
seasonal weather on efficiency of water use by corn. Agron. J. *49*, 362–
365.

MANGELSDORF, P. C., MACNEISH, R. S., and GALINAT, W. C. 1964.
Domestication of corn. Science *143*, 538–545.

MILLAR, C. E., and TURK, L. M. 1951. Fundamentals of Soil Science,
2nd Edition. John Wiley & Sons, New York.

NELSON, G. H., TALLEY, L. E., and ARONOVSKY, S. I. 1950. Chemical
composition of grain and seed hulls, nut shells, and fruit pits. Trans.
Am. Assoc. Cereal Chemists *8*, 58–68.

NELSON, O. E., JR. 1955. Purdue hybrid performance tests for 1955.
Popcorn Merchandiser *10*, No. 3, 3–9.

PETTY, H. B., and APPLE, J. W. 1967. Insects. *In* Advances in Corn
Production. Iowa State Univ. Press, Ames, Iowa.

PRINCE, A. B. 1954. Effects of nitrogen fertilization, plant spacing, and
variety on the protein composition of corn. Agron. J. *46*, 185–186.

QUACKENBUSH, F. W. *et al.* 1961A. Analysis of carotenoids in corn grain.
J. Agr. Food Chem. *9*, 132–135.

QUACKENBUSH, F. W. *et al.* 1961. Composition of corn. Analysis of
carotenoids in corn grain. J. Agr. Food Chem. *9*, 132–135.

QUACKENBUSH, F. W., FIRCH, J. G., BRUNSON, A. M., and HOUSE, L. R.
1963. Carotenoid, oil, and tocopherol content of corn inbreds. Cereal
Chem. *40*, 250–259.

RICHARDSON, D. L. 1957. Purdue hybrid performance trials encourag-
ing. Popcorn and Concessions Merchandiser *12*, No. 4, 10–17.

RICHARDSON, D. L. 1958. Two factors of early harvesting contribute
to popcorn quality. Concessionaire Merchandiser *13*, No. 4, 12–13.

RICHER, A. C. 1950. Basic teachings of Jordan plots easily translated
into useful farm practice. Penn. Agr. Expt. Sta. Bull. *515*.

RUSSELL, M. B., and DANIELSON, R. E. 1956. Time and depth patterns
of water use by corn. Agron J. *48*, 163–165.

SATHER, L. A., and CALVIN, L. D. 1963. Relation between preference
scores and objective and subjective quality measurements of canned
corn and pears. Food Technol. *17*, 917–920.

SHRADER, W. D., and PIERRE, J. J. 1967. Soil suitability and cropping
systems. *In* Advances in Corn Production. Iowa State Univ. Press,
Ames, Iowa.

STANIFORTH, D. W. 1957. Effects of annual grass weeds on the yield of
corn. Agron. J. *49*, 551–555.

THOMAS, W. 1956. Effect of plant population and rates of fertilizer
nitrogen on average weight of ears and yield of corn in the South.
Agron. J. *48*, 228–230.

TRESSLER, D. K., VAN ARSDEL, W. B., COPLEY, M. J., and WOOLRICH,
W. R. 1968. The Freezing Preservation of Food, 4th Edition. Avi
Publishing Co., Westport, Conn.

TRUSWELL, A. S., and BROCK, J. F. 1961. Effects of amino acid supplements on the nutritive value of maize protein for human adults. Am.
J. Clinical Nutr. *9*, 715–720.

ULLSTRUP, A. J. 1967. Diseases of corn and their control. *In* Advances in Corn Production. Iowa State Univ. Press, Ames, Iowa.

VENGRIS, J., COLBY, W. G., and DRAKE, M. 1955. Plant nutrient competition between weeds and corn. Agron. J. *47*, 213–216.

WALLACE, H. A., and BROWN, W. L. 1956. Corn and Its Early Fathers. Michigan State Univ. Press, East Lansing, Mich.

WATSON, S. A. 1967. Manufacture of corn and milo starches. *In* Starch Chemistry and Technology. Vol. II, Industrial Aspects. R. L. Whistler, and E. F. Paschall (Editors). Academic Press, New York.

WEATHERWAX, P. 1954. Indian Corn in Old America. Macmillan Co., New York.

WEI, L. S., STEINBERG, M. P., and NELSON, A. I. 1967. Quality of sweet corn during maturation as determined by two objective methods. Food Technol. *21*, 106A–108A.

WILLIS, W. O., LARSON, W. E., and KIRKHAM, H. 1957. Corn growth as affected by soil temperature and mulch. Agron. J. *49*, 323–328.

WOLF, M. J., BUZAN, C. L., MACMASTERS, M. M., and RIST, C. E. 1952A. Structure of the mature corn kernel. Gross anatomy and structural relationships. Cereal Chem. *29*, 321–333.

WOLF, M. J., BUZAN, C. L., MACMASTERS, M. M., and RIST, C. E. 1952B. Structure of the mature corn kernel. II. Microscopic structure of pericarp, seed coat, and hilar layer of dent corn. Cereal Chem. *29*, 334–348.

WOLF, M. J., BUZAN, C. L., MACMASTERS, M. M., and RIST, C. E. 1952C. Structure of the mature corn kernel. III. Microscopic structure of the endosperm of dent corn. Cereal Chem. *29*, 349–361.

WOLF, M. J., BUZAN, C. L., MACMASTERS, M. M., and RIST, C. E. 1952D. Structure of the mature corn kernel. IV. Microscopic structure of the germ of dent corn. Cereal Chem. *29*, 362–381.

WOLFE, M., and FOWDEN, L. 1957. Composition of the protein of whole maize seeds. Cereal Chem. *34*, 286–295.

ZUBER, M. S., SMITH, B. E., and GEHRKE, C. W. 1954. Crude protein of corn grain and stover as influenced by different hybrids, plant populations, and nitrogen levels. Agron. J. *46*, 257–261.

CHAPTER 3

Samuel A. Matz | **Oats**

ORIGIN AND IMPORTANCE

At the dawn of agriculture, a form of wild oats apparently existed as a weed in widely separated regions of the world. The primitive agriculturists, however, concentrated their efforts on the cultivation of the forerunners of modern wheat and barley. Oats continued to appear in the fields and was harvested with the more desirable grains. It made its way as an adventitious interloper in stocks of seeds of barley and emmer carried to other parts of the world from the cradle of civilization around the eastern end of the Mediterranean. At an unrecorded time during the historical era, it began to be grown in more or less pure stands, probably because it gave larger or more consistent yields than other grains grown under the climatic and soil conditions existing in some parts of northwestern Europe. Natural and controlled selection and perhaps some cross-breeding subsequently led to the development of modern types of oats. In the first century A.D., the Roman historian Pliny wrote that the Germanic peoples ate oats as a porridge. Preceding writers had referred to the plant as a weed, forage, or medicinal plant.

Some Russian authors have contended for a far Eastern origin of the common cultivated types, basing this opinion primarily on the pattern of distribution of certain kinds of wild oats.

Shortly after 1600, oats were brought to North America and planted by the early colonists. They were not extensively grown by the early settlers, probably because corn could serve many of the same functions as oats and gave considerably greater yields.

Today, the annual world oats crop probably exceeds four billion bushels taken from about 130,000,000 acres, making it fourth in importance among the cereals. Between 40 and 45% of the total harvest is produced in the United States and Canada. Other leading oat producing countries are Russia, Germany, and France. The grain is not nearly as important in commerce as wheat and corn. It is generally consumed in the locality, and frequently on the same farm, where it is produced. The chief use in most countries is as animal feed, with human consumption second, and industrial usage a distant third. The export trade is rather limited, ranging between 3 and 30 million bushels from the United States.

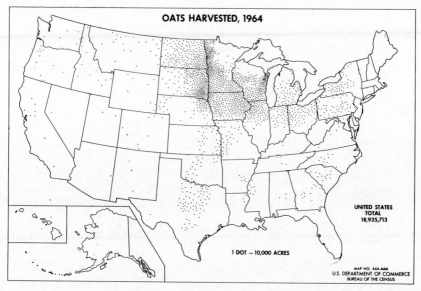

Courtesy of US Dept. Commerce

FIG. 19.   ACREAGE OF OATS HARVESTED IN THE UNITED STATES

BOTANY OF THE OAT PLANT[1]

The Genus Avena

The oat plant is classified as follows:

Phylum:	Spermatophyta
Class:	Angiospermae
Subclass:	Monocotyledones
Order:	Graminales
Family:	Poaceae (Graminaceae)
Tribe:	Aveneae
Genus:	Avena

There are annual and perennial oats, but all cultivated varieties are annuals and only the latter types will be discussed in any detail.

The genus *Avena* apparently was established in 1700 by Tournefort, a French explorer and botanist.   Most species of oats known today were described as early as 1750 by Linnaeus, the great Swedish botanist.

As previously indicated, the common oat (*Avena sativa*) is grown in the cooler and moister regions of the temperate zones.   This species

---

[1] Much of this section was written by Dr. T. R. Stanton.

constitutes most of the oats produced today. The red oat, *A. byzantina*, is grown in regions considered too warm for satisfactory growth of the common oat. If it were not for these heat-tolerant red oats, production of the grain would be much less important in the southern United States, South America, Australia, and the Mediterranean countries of Europe.

Stanton (1955) published a description of the oat plant as follows:

"The oat plant is an annual grass belonging to the genus *Avena*. Cultivated oats are derived chiefly from two species, the common wild oat (*A. fatua* L.) and the wild red oat (*A. sterilis* L.). The principal derivatives of the former are the common oat (*A. sativa* L.), including the side oat (*A. orientalis* Schreb.). Of the latter, the only important cultivated form is *A. byzantina* (C. Koch), including *A. sterilis algeriensis* Trabut.

"Under average conditions the oat plant produces from 3 to 5 hollow stems, or culms, varying from $1/8$ to $1/4$ in. in diameter and from 2 to 5 ft in height. The roots are small, numerous, and fibrous, and penetrate the soil to a depth of several feet. The leaves average about 10 in. in length and $5/8$ in. in width. The panicles, or heads, are either spreading (equilateral, or tree-like) or one-sided (unilateral, horse-mane, or banner-like). By far the greater number of cultivated varieties have spreading panicles. The grain is produced on small branches, in spikelets, varying in number from 20 to 150 per panicle. Each spikelet contains 2 or 3 florets or grains except those of the hull-less or naked oat which contain 4 to 8. The spikelet is loosely enclosed within the outer glumes (chaff). The kernels, except in the hull-less oat, are tightly enclosed in the lemmas or inner glumes and palea. The lemma or hull ranges in color from white, yellow, gray, and red, to black, and may be awned or awnless. The kernel, or more properly the caryopsis, without its adhering glumes, is very slender, ranging from $5/16$ to $7/16$ in. in length and from $1/16$ to $2/16$ in. in width. The kernel constitutes 65 to 75% of the total weight of the whole grain."

For additional information on the oat species and varieties, including description, history, and distribution of the species and varieties, see Stanton (1955).

Many Varieties.—Several hundred varieties of oats may be differentiated on the basis of botanical characters. Through the years the total number of named commercial varieties that appeared from time to time has run into the thousands. The so-called World Collection of Oats on file in the Field Crops Research Branch, Agricultural Research Service, US Dept. of Agr., Beltsville, Md., contains nearly 5,000 varieties and strains. Most of these are simply named or unnamed strains of a much smaller number of definite botanical varieties and types; that is, there are many duplicates. During the last two decades the adoption by farmers of improved or selected varieties recommended by State Agricultural Experiment Stations and Extension Services has, fortunately, reduced the number of named, often nondescript, disease-susceptible, low-yielding varieties that plagued the farmer for years.

Classification of Species and Varieties

Stanton (1953), in his rather comprehensive studies on the identification and classification of oats, described 12 species or subspecies of *Avena* including: (1) Large Naked Oat (*Avena nuda* L.), (2) Small Naked Oat (*A. nudibrevis* var., *A. nuda* L. ssp. *biaristata* (Alef.) Asch. & Graeb.), (3) Wild Red Oat (*A. Sterilis* L.), (4) Red Oat (*A. byzantina* C. Koch), (5) Desert Oat (*A. wiestii* Steud.), (6) Slender Oat (*A. barbata* Brot.), (7) Sand Oat (*A. strigosa* Schreb.), (8) Abyssinian Oat (*A. abyssinica* Hochst), (9) Short Oat (*A. brevis* Roth), (10) Wild Oat (*A. fatua* L.), (11) (Common) Tree Oat (*A. sativa* L. ssp. *diffusa* (Neils.) Asch. and Graeb.), and (12) (Common) Side Oat (*A. sativa* L. ssp. *orientalis* Schreb.).

Unfortunately, space permits brief discussion of only the varieties of Red, (Common) Tree, and (Common) Side Oats in this book. Brief reference to the large Naked or Hull-less Oat also is made due to the fact that a few hull-less oats are grown commercially on farms in the United States.

Stanton (1953) described only 4 varieties of the Large Naked Oat and named 6 synonymous varieties. Because of low yield and other agronomic disadvantages, the Naked Oat never attained much economic importance in the United States. A few are grown each year more as a novelty than as a standard crop by farmers who become curious.

Stanton described 50 varieties of the Red Oat. Of these, 28 have been grouped in the key as representing true sterilis types with second floret separating quite consistently by basifracture, and 22 as intermediate types with second florets not separating consistently by basifracture, heterofracture (intermediate), or disarticulation, the last being best exemplified by the Burt and Fulghum varieties. In addition, 60 synonymous varieties were named.

Stanton described 146 varieties of the (Common) Tree Oat and named 165 synonymous varieties. He also described 18 varieties of the (Common) Side Oat and named 16 synonymous varieties. As a group, most of the varieties with side panicles are of little economic importance because of their low yielding power and inferior grain quality. They also have low tillering capacities and large, thick hulled grains. One Side Oat variety, White Tartar (White Russian), is of value for breeding because of high resistance to certain races of stem rust.

Several other botanists, e.g., Etheridge and Marquand, have classified the varieties of oats. Etheridge (1916), in his classification studies based on 731 collections, established 55 botanical varieties including both the Red Oat and the Common Oat. The remaining strains were either classified as synonymous varieties, or were discarded as representing badly mixed nondescript oats of no value for classification.

Marquand (1922), an English botanist, in a classification of the species and varieties of oats mostly grown in the British Isles, recognized 112 varieties and subvarieties of Red and Common Oats. He also named a few synonymous varieties.

During recent years the breeding and distribution of many new disease-resistant varieties originating as selections mostly from crosses between Red and Common Oats has further complicated the problem of varietal identification and classification owing to similarity in many morphological characters and disease reactions.

Genetics and Inheritance in Oats.—Briefly, the basic genetic principles determining the mode of inheritance of morphological characters and disease reactions in oats are the same as for wheat and barley. Most characters are inherited on a monogenic basis, including the 3:1 and 1:2:1 segregations. A relatively smaller number of characters are inherited on a digenic or trigenic basis, including the 15:1 and 13:3 and 63:1 segregations. There remain many characters, mostly of a quantitative nature, that depend upon so-called multiple factors for their inheritance.

The chromosomes of the genus Avena fall into 3 groups characterized by the numbers $n = 7$, 14, and 21. The chromosome numbers compiled by O'Mara (1961) of some of the important species are shown in Table 21.

The gross plant characteristics which have received the attention of geneticists are: (1) height, (2) tillering, (3) lodging resistance, (4) plant colors, (5) panicle shape, (6) floret disjunction, (7) ligule type, and (8) maturity. The spikelet and floral characters which are inherited on a known basis are: (1) spikelet separation, (2) floret disjunction, (3) rachilla length, (4) awns, (5) pubescence, (6) lemma color, (7) waxy lemma, (8) luminescence of lemma, (9) hull-lessness

TABLE 21

IMPORTANT AVENA SPECIES ARRANGED BY NUMBERS OF CHROMOSOMES[1]

Diploids n = 7	Tetraploids n = 14	Hexaploids n = 21
clauda	barbata	fatua
pilosa	wiestii	sativa
longiglumis	vaviloviana	nuda
ventricosa	abyssinica	sterilis
strigosa		byzantina
		orientalis
		ludoviciana

[1] O'Mara (1961).

and the multiflorous spikelet, and (10) size and shape of kernel. The following physiologic factors have also been studied: (1) response to temperature, including winter hardiness, (2) germination, (3) yield, and (4) chemical composition. For further details consult Jensen (1961).

Atypical plants which have some characteristics of both the parental variety and A. fatua appear rather frequently in populations of cultivated oats. These "fatuoids" nearly always have the twisted and geniculate awns and sunken oval disarticulation surface surrounded by pubescence that are seen in wild oats. This is probably due to mutation although hybridization with A. fatua can produce the same effect. The mutation apparently affects a suppressor gene in the absence of which the fatuoid characteristics can be expressed.

A recent publication (Murphy et al. 1968) described a new tetraploid species (i.e., the 2n stage having 28 chromosomes) morphologically similar to the hexaploid A. sterilis. The new species, A. magna, has slightly larger florets and caryopsis, slightly thicker awns, and slightly longer pedicels, as compared to A. sterilis. As in the latter species, its spikelets articulate only at the base of the lowest floret, leaving a dissemination unit of three tightly attached florets. A. magna may come to have considerable economic importance because it has large groats with high protein content, and exhibits outstanding resistance to crown rust.

OAT CULTURE

Effect of Climate and Water Supply on Growth

Climate is the factor of primary importance governing oat growth. Oats are best suited to cool, moist climates, but the diversity of varietal types makes successful production possible over a range of climatic conditions wider than for any other cereal.

A small percentage of oat seeds will germinate at temperatures near freezing, and they require a comparatively cool temperature during the periods of germination and greatest growth. None of the oat varieties are as winter-hardy as some of the varieties of wheat and barley, however, and they do not withstand subzero temperatures for long periods in the absence of snow cover.

Coffman and Frey (1961) stated that the largest area of oat production in North America lies north of the line where average June temperatures are below 65°F, and that the area of winter oat culture lies to the south of the −5°F (December–February) isotherm. More generally, it can be said that oats are grown north of the 40th degree of latitude in North America, centering close to the 45° latitude, while in Europe and Asia the equivalent latitudes would be 50° and 52°. The southern limit of effective culture seems to be about 30° in the Northern Hemisphere although some cultivation does take place at relatively high altitudes south of this line. In the Southern Hemisphere, oats are grown for grain in New Zealand, Australia, and South Africa.

Oats apparently require more moisture to produce a given amount of dry matter than any other cereal except rice. Rainfall is frequently the limiting factor in yield. In North America, most oats are grown in an area receiving 18 in. or more of April to September precipitation, but substantial production is also achieved in the areas receiving only 15 in. Precipitation in August and September probably has no improving effect on yield except in the far northern growing areas. Coffman and Frey (1961), in their review of the influence of climate and physiological factors on growth of oats, summarized the earlier work on water requirements at various stages.

Soil and Nutrient Effects

Although there have been few studies on the relation of soil type to oats yield and quality, it is generally agreed that this grain is less selective than most of the other cereals. Grains in general grow best on Podzolic soils (forested soils of humid, temperate climates, including many areas of organic soils), Chernozemic soils (grass-covered soils of subhumid, semiarid, temperate climates) and, especially in southern North America, Latosolic soils (forested and savanna soils of humid areas). According to Coffman (1961B) oats in North America are generally grown on what are commonly called clay or sandy loams.

Nitrogen is often the limiting nutrient in production, especially on the less fertile and sandier soils. A good response is obtained from many of the nitrogenous fertilizers such as sodium nitrate, ammonium sulfate, anhydrous ammonia, urea, etc. They should be applied at seeding to promote vegetative growth and tillering. A preceding crop of clover will ordinarily supply enough nitrogen for excellent yields. Various authors report yields increase about 10 bu per acre for each 12 to 16 lb per acre of nitrogen added to the soil, when this nutrient is the limiting factor in growth. The top limit for addition will depend upon the soil type and other conditions. Severe lodging may result when the application of nitrogen exceeds the optimum rate (Shands and Chapman 1961).

Supplies of phosphates, potassium, and calcium are needed for oats, as for other crops. This plant is apparently able to thrive at lower levels of these nutrients than are corn, wheat, or some other cereals. Generally, the response to application of phosphorus is not as noticeable as that to nitrogen. The use of a soluble phosphate fertilizer at seeding time will usually result in a favorable response in fall oats (Shands and Chapman 1961). The amount of potassium required for high yields is less than for many other crops, but annual applications in the form of a complete fertilizer at planting time are recommended. Most soils do not benefit appreciably from lime, and excessive additions, resulting in a pH of 6.5 or more, may cause a manganese deficiency.

Oats are subject to retarded or defective growth on manganese-deficient soils. Use of 100 lb of manganese sulfate per acre, or the application of manganese to the foilage by spraying are effective remedies. Iron, copper, zinc, boron, and molybdenum are other essential minor nutrients for oats (Robinson and Edington 1945).

All varieties of oats grow slowly or not at all in saline soils. The harmful concentration seems to be around 1,000 ppm chloride.

Insects, Diseases, and Pathological Conditions

As with other crops, oats are subject to insect or mite attack from the time of seeding until the grain is consumed as food or feed. There are few insects which are specific to oats. Soil insects having larvae which feed on roots or underground portions of the stems of oats include wireworms (Elateridae), false wireworms (Tenebrionidae), and white grubs (Scarabaeidae). The predominant insects that attack oats above ground, feeding on the leaves, panicles, and stems, or sucking sap from the plant, are grasshoppers (Acrididae), cutworms (Phalaenidae), armyworms (*Pseudaletia unipuncta* (Haw.)), aphids (Aphidae), chinch bugs (*Blissus leucopterus* (Say)), leafhoppers (Cicadellidae), thrips (Thysanoptera), and frit flies

(Chloropidae).   The winter grain mite *(Penthaleus major* (Duges)) is the most important mite damaging oats (Dahms 1961).   The army worm is probably the most destructive of these pests, attacking oats in the late boot or headed stage.

Oats are more subject to lowered yield from disease than are other small grains.   Crown rust due to the fungus *Puccinia coronata* Cda., is probably the most widespread and destructive of these diseases (Simons and Murphy 1961).   It is most serious in humid areas. The other important rust disease, stem rust, is caused by the basidiomycete *Puccinia graminis* Pers. f. sp. *avenae* Erikss. & E. Henn. The smut diseases, loose smut and covered smut, due to *Ustilago avenae* (Pers.) Rostr. and *Ustilago kolleri* Wille., respectively, are also widespread and common.

There are also diseases of oats caused by viruses.   Oat pseudorosette, wheat mosaic, oat mosaic, barley yellow, and blue dwarf are among the known viruses.   Several bacteria are capable of damaging oats.   Halo blight is caused by *Pseudomonas coronafaciens* (Elliot), bacterial stripe blight by *Pseudomonas striafaciens* (Elliot), which is very similar to *P. coronafaciens*, and bacterial blight by *Xanthomonas translucens* (L. R. Jones, A. G. Johnson & Reddy).   Fungi other than rust and smut fungi which parasitize oats include various pythium species responsible for root necrosis, *Sclerospora macrospora* Sacc. which causes downey mildew, *Septoria avenae* Frank, and *Erysiphe graminis*, the agent of powdery mildew.   Several species of the genus Helminthosporum also attack oats.

Nematodes are destructive to oats in some areas of Europe but are apparently of considerably less importance in North America.

There are economically important diseases which are not due to attack on the plant by another organism.   These include trace element deficiencies, particularly relatable to a lack of adequate manganese or copper, and blast, which is apparently due to extremes of environmental conditions damaging the plant at a critical stage in its development.   Susceptibility to blast is apparently related to genetic constitution.

A complete review on oat diseases has been published by Simons and Murphy (1961).

Harvesting and Storage

In the United States, harvesting begins in May in the most southerly parts, and continues into September in northern United States and Canada.   The greatest amount of oats is harvested in July and August.   Unusual weather conditions may cause the crop to be left standing until considerably later in the year.

Most oats harvested for grain in the United States is combined. The much greater labor efficiency resulting from combine use may at times be offset by loss of grain or deterioration due to excessive moisture in the grain.

As little as 12% of moisture is said to be too high for safe storage in warmer climates, but the usual level of 14 to 15% quoted for several other grains is probably satisfactory in most cases.   The effects of excessive moisture, such as heating, discoloration, loss of viability, and development of visible mycelia and an off-odor, are the same as for other cereal grains.   These difficulties can be forestalled by artificial drying procedures such as blowing hot air through the bin. Oats are less susceptible than wheat and corn to heat damage during such drying, except for the effects on viability.

QUALITY FACTORS IN OATS

Nearly all trade in cereal grains in the United States is based on the Official Grain Standards (Anon. 1964).   Under these Standards, grain offered for sale as "Oats" must contain at least 75% of *A. sativa*, *A. byzantina*, or a mixture of these species.   Oats are divided into the five classes of white oats, gray oats, red oats, black oats, and mixed oats.   Generally speaking, the standards allow up to 10% (by weight) of oats of colors other than the one specified in the class.

TABLE 22

GRADE REQUIREMENTS FOR OATS[1]

Grade	Minimum Limits		Maximum Limits		
	Test Weight per Bushel (Lb)	Sound Cultivated Oats (%)	Heat-damaged Kernels (%)	Foreign Material (%)	Wild Oats (%)
1[2]	34	97	0.1	2.0	2.0
2[3]	32	94	0.3	3.0	3.0
3[4]	30	90	1.0	4.0	5.0
4[5]	27	80	3.0	5.0	10.0
Sample[6]	...	...	...	...	...

[1] Anon. 1964.
[2] The oats in grade No. 1 White Oats may contain not more than 5.0% of red oats, gray oats, and black oats, singly or in combination, of which not more than 2.0% may be black oats.
[3] The oats in grade No. 2 White Oats may contain not more than 3.0% of black oats.
[4] Oats that are slightly weathered shall be graded not higher than No. 3.
[5] Oats that are badly stained or materially weathered shall be graded not higher than No. 4.
[6] Sample grade shall be oats which do not meet the requirements for any of the grades from No. 1 to No. 4, inclusive; or which contain more than 16.0% of moisture; or which contain stones; or which are musty, or sour, or heating; or which have any commercially objectionable foreign odor except of must or garlic; or which are otherwise of distinctly low grade.

Each class is further subdivided into grades which are based on the characteristics shown in Table 22.

Oats to be milled for human consumption should be plump, sound, and free from heat damage, foreign odors, wild onion seed, smut, must, and molds. Usually only grade 1 or grade 2 oats are employed. Plumpness contributes to a good yield and desirable texture in the finished oatmeal. Heat or mold damage, or the presence of foreign materials leads to off-color particles and off-flavors. There is no selection of lots on the basis of protein or other nutrient content.

STRUCTURE AND COMPOSITION OF OATS

Cereal grains differ from other grasses in having simple, dry, and indehiscent fruits. The pericarp or fruit coat is attached to the remainder of the fruit about the entire periphery. In the commercially important hulled oats, the caryopsis is inclosed in a covering made up of floral envelopes. This is missing in the seldom-grown naked or hull-less oats.

The gross physical structure of the oat groat is similar to that of the kernels of wheat and barley, however, it is covered with numerous trichomes or hairlike protuberances. The three major divisions into which the groat can be seen to be divided are the bran, endosperm, and germ. Starting from the outside of the groat, the bran layer consists of the epidermis, seed coat, hyaline layer, and aleurone

TABLE 23

COMPOSITION OF OATS, GROATS, AND MEAL[1]

Component (Dry Weight Basis)	Dry Milling Oats	Finished Groats	Oat Meal
Moisture, %	7.5	7.5	8.1
Crude protein, % (N × 6.25)	13.0	17.0	16.8
Crude fat, %	5.5	7.7	6.7
Crude fiber, %	11.8	1.6	3.9
Ash, %	3.7	2.0	2.2
Nitrogen free extract, %	66.0	71.6	70.5
Iron, ppm	51	47	51
Calcium, ppm	700	570	580
Phosphorus, %	0.367	0.501	0.445
Manganese, ppm	34	34	48
Thiamine, mg/100 gm	0.65	0.77	0.78
Riboflavin, mg/100 gm	0.14	0.14	0.17
Niacin, mg/100 gm	1.15	0.97	1.25
Pantothenic acid, mg/100 gm	0.86	1.36	1.19
Folic acid, mg/100 gm	0.034	0.057	0.06
Choline chloride, mg/100 gm	107	120	120
Pyridoxine, mg/100 gm	0.20	0.12	...

[1] Adapted from "Facts on Oats," Quaker Oats Co.

cells, in that order.   The germ is made up of embryonic cells and cell
wall tissues, while the endosperm is composed mainly of starch cells.

The oat germ, when viewed in a longitudinal section, extends
roughly $1/3$ of the way up the ventral side of the groat.   It is larger
and narrower than the germ of wheat.   Bounding the germ is a layer
of columnar cells which are strongly colored by protein strains.

Compared to other cereals, oat groats are characterized by low
carbohydrate content, and higher protein and fat content.   Using
selective stains followed by microscopic observation of sections,
House and Werly (1960) identified lignin in the epidermis, aleurone
cell walls, endosperm cell walls, and germ cell walls.   The epidermis
yielded the most positive test.   Cellulose was found in the hyaline
layers, aleurone cell walls, endosperm cell walls, and germ cell walls.
Protein would be expected to be more or less ubiquitous, but positive
reactions for it were found in the aleurone cells, germ cells, and endosperm cell walls.   Starch appeared to be confined to the endosperm,
where it occurred in well-defined aggregates.   The germ and the
aleurone cells were determined to be the major lipid depots.   There
were traces of pectin in the horny endosperm and epidermis.

According to data on file in the Quaker Oats Research Laboratories,
the percentages of ash, fiber, and nitrogen-free extract vary only
slightly among samples of oat grain taken from many different places
throughout the world.   The different streams of oats separated before milling are also similar in composition for most samples, but
there is a considerable variation in the composition of samples if they
are widely different in hull percentage.

Carbohydrates

Whole oats, including hulls, will yield about 65% nitrogen-free
extract, on a dry basis.   Starch and other carbohydrate polymers
make up about 90% of this material.   Reducing sugars in the extract are quite low, usually less than 0.1%, while total sugars are
often near 1.4%.   Whole oats contain about 14% pentosans, mainly
araban and xylan.   The highest concentration of pentosans is in the
hulls although the groats will have around 4%.   Their content of
pentosans has made oat hulls an important raw material for the manufacture of furfural, a chemical intermediate and solvent.

The fiber of oats, found mainly in the hulls, consists principally of
cellulose, hemicellulose, and lignin.   According to Nelson *et al.*
(1950), dry oat hulls contain about 16.7% lignin and 29.4% alpha-
cellulose.

Starch.—Between 33 and 43% starch (on a dry basis) can be extracted from whole oats, and the actual starch content is undoubtedly somewhat higher (MacMasters *et al.* 1947). The granules are very irregular in shape, often assuming a polyhedral configuration, although ovoid and hemispherical particles are also seen in quantity. An occasional spindle-shaped or pointed oval granule will be seen; these are characteristic of oats. The average size varies according to the location within the kernel, the largest and most loosely packed granules being found in the cheek near the crease, while the horny endosperm contains smaller particles. Most granules will fall within the size range of 3 to 10 μ in their largest dimension, giving oats the smallest starch particle of all the cereal grains except rice. The hilum is faint or invisible, and lamellae are rarely seen.

The individual granules develop in bundles or clusters about 60 μ in diameter, which fill most of the central space within the endosperm cells. These compound clusters are usually broken up during processing of the kernel, yielding the irregularly shaped granules seen in most oat starch preparations.

Birefringence of the granule is relatively weak, with the polarization cross being centrally located in most cases. The amylose content is about 23 or 24% (Deatherage *et al.* 1955). Apparently, there are no "waxy" oat varieties.

Lipids of Oats

The ether or petroleum ether extract of oats can vary from about 4 to about 10%, and is distributed among the parts of the grain as shown in Table 24, based on the analyses of many examples of commercial grain by the Quaker Oats Co. The amount of lipid extracted depends upon the solvent, of course. When the amount extracted by diethyl ether is taken as 100, the equivalent figures for material extracted by other common solvents are: petroleum ether 97, carbon tetrachloride 104, chloroform 110, benzene 113, acetone 113, and ethanol 128. The ethanol and acetone extractables values are in the same range as the figures obtained when fat is determined by acid hydrolysis methods.

Oat oil is nearly neutral in reaction when first extracted, but soon develops free fatty acids under common conditions of storage. Analytical data considered to be reasonably representative of normal oat oil samples are given in Table 25.

Oat oil contains 42 to 45% polyunsaturated fats. Oats is an excellent source of linoleic acid, an essential fatty acid for the human diet.

Solidification point of oat oil samples varies from 41° to about 68°F.

TABLE 24

DISTRIBUTION OF FAT CONTENT OF OATS[1]

	Ether Extract, Dry Basis (%)
Whole grain	5.4
Groats	7.6
Hulls	0.62
Germ (hand-dissected)	11.2
Germ-free groats	5.8
Germ plus bran coats	7.4–9.0
Endosperm	6.2–6.7

[1] Adapted from data in "Facts on Oats" (1950).

TABLE 25

CHARACTERISTICS OF OAT OIL

	US Oats[1]	Japanese Oats[2]
Specific gravity at 77°F	0.9191	. . .
Refractive index at 68°F	1.4710	. . .
Saponification value	190.6	196.1
Wijs iodine number	107.2	114.7
Unsaponifiable matter, %	2.26	1.74
Fatty acids from oil		
Palmitic	10%[3]	9.72%
Oleic	58.5%[3]	52.4%
α-linoleic	17.2%[3]	17.4%
β-linoleic	13.9%[3]	

[1] Munro and Binnington (1928).
[2] Takahashi et al. (1935); an average of two types of oats given.
[3] Amberger and Wheeler-Hill (1927).

TABLE 26

AMINO ACID COMPOSITION OF OAT PROTEINS[1]

Amino Acid	Mean	Range	Number of References
Alanine	6.2	. . .	1
Arginine	6.4	5.9–6.8	4
Aspartic acid	3.7	. . .	1
Cystine	1.8	. . .	2
Glutamic acid	20.0	. . .	1
Glycine	2.4	0.8–4.1	2
Histidine	2.2	1.3–2.3	4
Isoleucine	4.8	4.2–5.1	4
Leucine	7.0	5.7–8.0	4
Lysine	3.4	3.3–3.6	4
Methionine	1.6	1.2–2.0	4
Phenylalanine	5.0	4.7–5.5	4
Proline	5.7	. . .	1
Serine	4.5	. . .	1
Threonine	3.3	3.0–3.6	4
Tryptophan	1.3	1.2–1.3	3
Tyrosine	3.8	3.2–4.5	2
Valine	5.8	5.4–6.6	4

[1] Adapted from a similar table in "Facts on Oats" (1950). Data from Booth et al. (1951), Heathcote (1950), McElroy et al. (1949), Block and Bolling (1951), and Carroll et al. (1949). Values are based on 16% nitrogen, dry weight.

Nitrogenous Substances

Various references indicate that from 85 to 94% of the nitrogen in oats is present in amide or amino groups. The usual data presented as protein in oats are the Kjeldahl-determined nitrogen contents multiplied by 6.25. This factor is not universally accepted. When so calculated, the protein of a typical example of milling oats will average about 13.1%, hulls about 4.5%, and finished groats about 16.9% (dry basis). Table 26 summarizes data on amino acid composition of oat proteins taken from several recent publications.

BIBLIOGRAPHY

ALSTON, A. M. 1966. Effects of limestone, nitrogen, potassium, and magnesium fertilizers on magnesium absorption by oats and barley. J. Agr. Sci. *67*, 1–6.

AMBERGER, K., and WHEELER-HILL, E. 1927. The composition of oat oil. Z. Untersuch. Lebensm. *54*, 417–437.

ANON. 1964. Official Grain Standards of the United States. US Dept. Agr. SRA-AMS *177*.

AVERY, G. S., JR. 1930. Comparative anatomy and morphology of embryos and seedlings of maize, oats, and wheat. Botan. Gaz. *89*, 1–39.

BAKER, D., NEUSTADT, M. H., and ZELENY, L. 1957. Application of the fat acidity test as an index of grain deterioration. Cereal Chem. *34*, 226–233.

BLOCK, R. J., and BOLLING, B. 1951. The Amino Acid Composition of Proteins and Food. 2nd Edition. Charles C Thomas, Springfield, Ill.

BONNETT, O. T. 1961. Morphology and development. *In* Oats and Oat Improvement. F. A. COFFMAN (Editor). American Society of Agronomy, Madison, Wisc.

BOOTH, R. R., HARE, J. H., and VAN LANDINGHAM, A. H. 1951. The glycine content of poultry ration components. Proc. W. Va. Acad. Sci. *23*, 69–75.

BRIAN, R. C. 1967. Action of plant growth regulators. IV. Adsorption of unsubstituted and 2,6-dichloro-aromatic acids to oat monolayers. Plant Physiol. *42*, 1197–1201.

BROWNLEE, H. J., and GUNDERSON, F. L. 1938. Oats and oat products— culture, botany, seed structure, milling, composition, and uses. Cereal Chem. *15*, 257–272.

CARROLL, R. W., HENSLEY, G. W., and PETERS, F. N., JR. 1949. Unpublished data. Quaker Oats Co., Barrington, Ill.

CHESHIRE, M. V., DE KOCK, P. C., and INKSON, R. H. E. 1967. Factors affecting the copper content of oats grown in peat. J. Sci. Food Agr. *18*, 156–160.

COFFMAN, F. A. 1946. Origin of cultivated oats. J. Am. Soc. Agron. *38*, 983–1002.

COFFMAN, F. A. 1961A. Origin and history. *In* Oats and Oat Improvement. F. A. COFFMAN (Editor). American Society of Agronomy, Madison, Wisc.

COFFMAN, F. A. 1961B. World importance and distribution. *In* Oats and Oat Improvement. F. A. COFFMAN (Editor). American Society of Agronomy, Madison, Wisc.

COFFMAN, F. A., and FREY, K. J. 1961. Influence of climate and physiologic factors on growth in oats. *In* Oats and Oat Improvement. F. A. COFFMAN (Editor). American Society of Agronomy, Madison, Wisc.

COFFMAN, F. A., MURPHY, H. C., and CHAPMAN, W. H. 1961. Oat breeding. *In* Oats and Oat Improvement. F. A. COFFMAN (Editor). American Society of Agronomy, Madison, Wisc.

COLLIER, G. A. 1949. Grain production and marketing. US Dept. Agr. Prod. Marketing Admin. Misc. Publ. *692.*

CUNNINGHAM, D. K., GEDDES, W. F., and ANDERSON, J. A. 1955A. Precipitation by various salts of the proteins extracted by formic acid from wheat, barley, rye, and oat flours. Cereal Chem. *32,* 192–199.

CUNNINGHAM, D. K., GEDDES, W. F., and ANDERSON, J. A. 1955B. Preparation and chemical characteristics of the cohesive proteins of wheat, barley, rye, and oats. Cereal Chem. *32,* 91–106.

DAHMS, R. G. 1961. Insects and mites that attack oats. *In* Oats and Oat Improvement. F. A. COFFMAN (Editor). American Society of Agronomy, Madison, Wisc.

DANIELS, D. G. H., KING, H. G. C., and MARTIN, H. F. 1963. Antioxidants in oats: esters of phenolic acids. J. Sci. Food Agr. *14,* 385–390.

DEATHERAGE, W. L., MACMASTERS, M. M., and RIST, C. E. 1955. A partial survey of amylose content in starch from domestic and foreign varieties of corn, wheat, and sorghum and from other starch-bearing plants. Trans. Am. Assoc. Cereal Chemists *13,* 31–42.

ETHERIDGE, W. C. 1916. A classification of the varieties of cultivated oats. Cornell Agr. Expt. Sta. Mem. *10,* 79–172.

FINDLAY, W. M. 1956. Oats: Their Cultivation and Use from Ancient Times to the Present Day. Oliver and Boyd, Edinburgh.

FREY, J. K. 1967. Mass selection for seed width in oat populations. Euphytica *16,* 341–349.

FREY, J. K., and CALDWELL, R. M. 1961. Oat breeding and pathologic techniques. *In* Oats and Oat Improvement. F. A. COFFMAN (Editor). American Society of Agronomy, Madison, Wisc.

FRIEDEMANN, T. E., and WITT, N. F. 1967. Determination of starch and soluble carbohydrates. II. Collaborative study of starch determination in cereal grains and cereal products. J. Assoc. Offic. Agr. Chemists *50,* 958–962.

FRIEDEMANN, T. E., WITT, N. F., and NEIGHBORS, B. W. 1967. Determination of starch and soluble carbohydrates. I. Development of method for grains, stock feeds, cereal foods, fruits, and vegetables. J. Assoc. Offic. Agr. Chemists *50,* 944–957.

FULTON, J. M., and FINDLAY, W. I. 1966. Influence of soil moisture and ambient temperature on the nutrient percentage of oat tissue. Can. J. Soil Sci. *46,* 75–81.

GROGG, B., and CALDWELL, E. F. 1958. Gelatinization of starchy materials in the farinograph. Cereal Chem. *35,* 196–200.

GUENZI, W. D., and MC CALLA, T. M. 1966. Phenolic acids in oats, wheat, sorghum, and maize residues and their phytotoxicity. Agron. J. *58*, 303–304.

HAINSWORTH, R. G., BAKER, O. E., and BRODELL, A. M. 1942. Seed-time and harvest today. US Dept. Agr. Misc. Publ. *485*, 1–97.

HAMILTON, A. 1966. Effects of nitrogenous and potassic salts with phosphates on yield and phosphorus, potassium and manganese contents of oats. Proc. Soil Sci. Soc. Am. *30*, 239–242.

HEATHCOTE, J. G. 1950. The protein quality of oats. Brit. J. Nutr. *4*, 145–154.

HOUSE, W. B., and WERLY, E. F. 1960. Development of an instant oat breakfast food. Midwest Research Institute Report to Quaker Oats Co. Unpublished.

HUTCHINSON, J. B., and MARTIN, H. F. 1955A. The chemical composition of oats. I. The oil and free fatty acid content of oats and groats. J. Agr. Sci. *45*, 411–427.

HUTCHINSON, J. B., and MARTIN, H. F. 1955B. The chemical composition of oats. II. The nitrogen content of oats and groats. J. Agr. Sci. *45*, 419–427.

JENSEN, N. F. 1961. Genetics and inheritance in oats. *In* Oats and Oat Improvement. F. A. COFFMAN (Editor). American Society of Agronomy, Madison, Wisc.

JOHNSON, A. A. 1961. Oat seed production and distribution. *In* Oats and Oat Improvement. F. A. COFFMAN (Editor). American Society of Agronomy, Madison, Wisc.

JONES, I. T., and HAYES, J. D. 1967. The effect of seed rate and growing season on flour oat cultivars. I. Grain yield and its components. J. Agr. Sci. *69*, 103–110.

JONES, L. H. P., and MILNE, A. A. 1963. Silica in the oat plant. I. Chemical and physical properties of the silica. Plant and Soil *18*, 207–220.

JONES, L. H. P., MILNE, A. A., and WADHAM, S. M. 1963. Silica in the oat plant. II. Distribution of silica in the plant. Plant and Soil *18*, 358–371.

KENT, N. L. 1957. Recent research on oatmeal. Cereal Sci. Today *2*, 83–91.

KIES, C., FOX, H. M., and WILLIAMS, E. R. 1967. Effect of nonspecific nitrogen supplementation on minimum corn protein requirement and first-limiting amino acid for adult men. J. Nutr. *92*, 377–383.

KLEPPER, B. 1967. Effects of osmotic pressure on exudation from corn roots. Australian J. Biol. Sci. *20*, 723–734.

KNIGHTS, B. A., and LAURIE, W. 1967. Application of combined gas-liquid chromatography-mass spectrometry to the identification of sterols in oat seed. Phytochem. *6*, 407–416.

MACMASTERS, M. M., SLOTTER, R. L., and JAEGER, C. M. 1947. The possible use of oats and other small grains for starch production. American Miller *75*, 82–83.

MARQUAND, C. V. B. 1922. Varieties of oats in cultivation. Univ. Col. Wales, Welsh Plant Breeding Sta. Series C, No. 2.

MCELROY, L. W., CLANDANIN, D. R., LOBAY, W., and PETHYBRIDGE, S. I. 1949. Nine essential amino acids in pure varieties of wheat, barley and oats. J. Nutr. *37*, 329–336.

MEIENHOFER, J. 1963. Oat protein. Z. Lebensmitt. Untersuch. *119*, 310–318.

MILTHORPE, F. L., and IVINS, J. D. 1966. The Growth of Cereals and Grasses. Butterworths, London.

MORGAN, D. E. 1967. Variations in the composition of oats and barley grown in Wales: Proximate constituents, available carbohydrates and "1,000 grain weights." J. Sci. Food Agr. *18*, 21–24.

MUNNS, D. N., JOHNSON, C. M., and JACOBSON, L. 1963A. Uptake and distribution of manganese in oat plants. I. Varietal variation. Plant and Soil *19*, 115–126.

MUNNS, D. N., JOHNSON, C. M., and JACOBSON, L. 1963B. Uptake and distribution of manganese in oat plants. III. Analysis of biotic and environmental effects. Plant and Soil *19*, 285–295.

MUNRO, L. A., and BINNINGTON, D. S. 1928. Extractor for the preparation of oat and other cereal oils. Ind. Eng. Chem. *20*, 425–427.

MURPHY, H. C., and COFFMAN, F. A. 1961. Genetics of disease resistance. *In* Oats and Oat Improvement. F. A. COFFMAN (Editor). American Society of Agronomy, Madison, Wisc.

MURPHY, H. C., SADANAGA, K., ZILLINSKY, F. J., TERRELL, E. E., and SMITH, R. T. 1968. *Avena magna:* An important new tetraploid species of oats. Science *159*, 103–104.

NELSON, G. H., TALLEY, L. E., and ARONOVSKY, S. I. 1950. Chemical composition of grain and seed hulls, nut shells, and fruit pits. Trans. Am. Assoc. Cereal Chemists *8*, 58–68.

O'MARA, J. G. 1961. Cytogenetics. *In* Oats and Oat Improvement. F. A. COFFMAN (Editor). American Society of Agronomy, Madison, Wisc.

POKORNÝ, J., ZEMAN, I., and JANÍČEK, G. 1961. Chemistry of oats and the products therefrom. V. Composition of saturated fatty acids of different fractions of oat grain. Sborn. Praz. Vrs. Skol. Chem.-Technol., Potravin. Technol. *5*, 351–365.

ROBINSON, W. O., and EDINGTON, G. 1945. Minor elements in plants and some accumulator species. Soil Sci. *60*, 15–28.

SAMPSON, D. R. 1954. On the origin of oats. Bot. Mus. Leaf., Harvard Univ., *16*, No. 10, 265–303.

SHANDS, H. L., and CHAPMAN, W. H. 1961. Culture and production of oats in North America. *In* Oats and Oat Improvement. F. A. COFFMAN (Editor). American Society of Agronomy, Madison, Wisc.

SIMONS, M. D., and MURPHY, H. C. 1961. Oat diseases. *In* Oats and Oat Improvement. F. A. COFFMAN (Editor). American Society of Agronomy, Madison, Wisc.

STANTON, T. R. 1937. Superior germplasm in oats. US Dept. of Agr. Yearbook Agr. *1937*, 347–414.

STANTON, T. R. 1953. Production, harvesting, processing, utilization, and economic importance of oats. Econ. Bot. *7*, 43–64.

STANTON, T. R. 1955. Oat identification and classification. US Dept. Agr. Tech. Bull. *1100*.

STANTON, T. R. 1959. Oats. *In* The Chemistry and Technology of Cereals as Food and Feed. S. A. MATZ (Editor). Avi Publishing Co., Westport, Conn.

STANTON, T. R. 1961. Classification of *Avena*. *In* Oats and Oat Improvement. F. A. COFFMAN (Editor). American Society of Agronomy, Madison, Wisc.

STEER, W. M., GUNNING, B. E. S., GRAHAM, T. A., and CARR, D. J. 1968. Isolation, properties, and structure of Fraction I Protein from *Avena sativa* L. Planta *79*, 254–267.

STOSKOPF, N. C., KLINCK, H. R., and STEPPLER, H. A. 1966. Temperature in relation to growth and net assimilation rate of oats. Can. J. Plant Sci. *46*, 397–404.

TAKAHASHI, E., TASE, T., and SAEGI, Y. 1935. Food chemistry of oats. V. Fats and oils of oats produced in Hokkaido. J. Agr. Chem. Soc. Japan *11*, 199–205.

THOMAS, J. W., and REDDY, B. S. 1962. Sprouted oats as feed for dairy cows. Mich. Agr. Expt. Sta. Quart. Bull. *44*, 654–665.

UDVARDY, J., FARKAS, G. L., MARRE, E., and FORTI, G. 1967. The effects of sucrose and light on the level of soluble and particle-bound ribonuclease activities in excised *Avena* leaves. Physiol. Plantarum *20*, 781–788.

WATSON, S. A. 1967. Manufacture of corn and milo starches. *In* Starch Chemistry and Technology. Vol. II. Industrial Aspects. R. L. WHISTLER and E. F. PASCHALL (Editors). Academic Press, New York.

WESTERN, D. E., and GRAHAM, W. R., JR. 1961. Marketing, processing, uses and composition of oats and oat products. *In* Oats and Oat Improvement. F. A. COFFMAN (Editor). American Society of Agronomy, Madison, Wisc.

WIGGINS, S. C., and FREY, K. J. 1958. The ratio of alcohol-soluble to total nitrogen in developing oat seeds. Cereal Chem. *35*, 235–239.

CHAPTER 4

Allan D. Dickson | **Barley**

INTRODUCTION

Origin of Cultivated Barley

Barley is one of the world's oldest domesticated crops, and probably shares with wheat the distinction of being the first wild plant form brought under cultivation.

The most likely progenitor of cultivated varieties is a wild species in the section Cerealia Ands. of the genus *Hordeum* which is found wild in areas of Southwestern Asia. *Hordeum spontaneum* C. Koch is a 2-rowed type with brittle rachis and was first described in 1848. A wild 6-rowed barley, *Hordeum agriocrithon* Aberg, was found by Aberg in 1938 and was considered to be the progenitor of cultivated 6-rowed barleys. This barley has a tough rachis and is not found in the wild away from areas of cultivated barleys. It has now been established that this barley was probably produced by natural crossing between *H. spontaneum* and domesticated 6-rowed types and could not have existed in the wild conditions.

During the last 30 yr, botanical and archaeological research has resulted in a feasible phylogenetic scheme for barley. *The Origin and Early History of Cultivated Barleys* by Clark (1967) reviews the early theories and, more importantly, the information based on more recent archaeological expeditions.

Barley was probably first used in agriculture in Western Asia, perhaps as early as 7000 BC. The earliest records of barley (radio carbon dating ±6750 BC) excavated at Jarmo in Iraqi-Kurdistan were similar to *H. spontaneum* in many ways, but had a tough rather than brittle rachis. Helbaek considers this as positive evidence that this was a cultivated rather than a wild form. According to Helbaek's monophyletic theory of origin, the 2-rowed type was taken to the southern alluvial plains by colonists, and a 6-rowed barley developed from it by mutation. Probably due to better adaptation

ALLAN D. DICKSON is Research Chemist, Barley and Malt Laboratory[1] Crops Research Division, Agricultural Research Service, US Dept. of Agriculture, Madison, Wisc.

[1] The work of the Barley and Malt Laboratory is supported in part by a research grant from the Malting Barley Improvement Association and is published with the permission of the Wisconsin Agricultural Experiment Station.

and higher yield, the 6-rowed types spread to Egypt and later to Europe and other areas. Except for a few areas, 2-rowed barleys did not again appear in quantity until well into the Christian era. The existing records do not indicate whether the later 2-rowed barleys were derived from existing 6-rowed forms by mutation, or whether they may have originated from the early Jarmo barley that was maintained and developed in certain limited areas as yet unknown.

Barley has been an important food plant during the development of agriculture and the grain was used as flour and to prepare fermented beverages. According to Weaver (1950) barley was first brought to the North American Continent for malting and the production of beer.

Importance in the Economy of the United States

Adaptability.—Barley is certainly the most widely adapted cereal crop and is possibly more widely adapted than any other cultivated crop. Its cultivation in Europe extends beyond the Arctic Circle, where it reaches 70°N latitude in Norway. To the south, barley culture extends within a few degrees of the Equator in the high mountain regions of Ethiopia, according to Nuttonson (1957). In the United States, barley is best adapted to the Northern and Western states. Over the long period of its cultivation, geographical environment and efforts of man have resulted in many diverse types. There are spring types that mature in 60 to 70 days and winter types that may require as long as 180 days. In general, barley is able to mature in a shorter season than any other major grain crop. Barley is not extremely winter hardy, nor is it favored by hot humid weather. It is the most salt tolerant of all cereals according to Spector (1956), and therefore best adapted to alkali soils. However, it is best adapted to and yields highest on fertile soils. Profitable barley production over long periods of time has been confined to areas with average summer temperatures of 70° F or less, annual precipitation not in excess of 35 in., and average relative humidity of less than 50%, according to Weaver (1950).

Economic Importance.—Barley is the major grain crop for feed and food in Northern areas of the world, or at high elevations where its short growing season makes it more dependable than wheat or oats. In the United States, it ranks fifth in total production, being exceeded by corn, wheat, oats, and sorghum. In Northern areas, where corn is not well adapted, it is the major feed grain.

Barley is by far the most important cereal grain for malting because of special physical and chemical properties. Approximately $1/4$ of the annual production, about 100 million bushels, is used for this purpose annually in the United States. Relatively small quantities of unmalted barley are used for food products in the United States, but moderate quantities are exported to Asiatic countries in some years.

BOTANY OF THE BARLEY PLANT

Description of Plant and Seed

Barley is one of the cereal members of the grass family. Winter and spring types exist. The plant consists of roots, leaves, stems, and flower parts. The grain is produced in spikes, or heads, at the top of the stems. Mature barley plants vary in height from 12 to 48 in., the height depending upon type or variety and growing conditions, but 30 in. has been given as a usual height by Shands and Dickson (1953). The flowers of the plant are arranged in spikelets on the head and are attached at nodes of a flat zigzag rachis, or central stem. Three spikelets, each with a pair of glumes, are attached at each node of the rachis and successive spikelets are located on alternate sides of the rachis. The spikelet is composed of the male and female flower parts, enclosed within the lemma and palea. The rachilla is attached to the germ end of the kernel and lies in the furrow of the palea. The lemma terminates in the awn, or beard, in awned varieties, but hooded barleys have short three-pronged appendages instead of awns. The florets open only for a short period at pollination, so barley is naturally self-pollinated.

The mature kernel is composed of the hulls (lemma and palea) enclosing the caryopsis, and of the rachilla. In most varieties, the hulls are cemented to the caryopsis and make up a part of the threshed kernel. In naked or hulless barleys, the kernel threshes free as does wheat. In 6-rowed barleys, the 3 spikelets at each node are fertile and produce kernels, while in 2-rowed forms only the central spikelet is fertile and develops a grain. Size, shape, and color of the kernels vary widely depending upon type and variety.

Classification of Cultivated Species

Plants of the genus *Hordeum* of the grass family, *Gramineae*, have simple spikes. The cultivated types of barley and the two most closely related wild forms are included in the section Cerealia. Wiebe and Reid (1961) classify the cultivated barleys into three species on the basis of brittleness of the rachis and number of kernel rows on the spike. They also give detailed descriptions of taxonomic characters for growth characteristics of the plant, spike, and kernel, and include keys for identification of varieties commonly grown at that time.   Plant and growth characteristics such as spring or winter, time of heading, hairiness of leaf sheaths, and collar shape at the node beneath the spike are relatively constant factors.   Spike characteristics such as stigma hairiness, number of rows of kernels, hairiness of rachis edges, length of outer glume awns, and nature of the lemma awn or other appendages are considered most useful in identifying varieties.   Kernel characteristics, evident in threshed samples, are also useful, and will be discussed briefly in a later section.

Most of the cultivated barleys have been classified into the two groups, *H. vulgare* L., the 6-rowed barleys, and *H. distichum* L., the 2-rowed types.   Typical examples are given in Fig. 20.

In the common 6-rowed barleys, 3 kernels develop at each rachis node.   The median kernel is slightly larger than the lateral kernels on each side and is symmetrical in shape.   The 2 lateral kernels are twisted and the twist is more pronounced at the attachment end of the kernel.   The Manchurian type 6-rowed barleys commonly grown

FIG. 20.   TYPICAL HEADS OF COMMONLY GROWN BARLEYS

(A) Six-rowed *Hordeum vulgare*.   (B) Two-rowed *Hordeum distichum*.

in the northcentral part of the United States and important as malting barleys are typical of *H. vulgare* L. and have medium-size kernels weighing approximately 36 mg. Varieties with rough or smooth awns, white or blue aleurone color, and nodding or erect heads are included.

Also included in this group are the coast-type barleys commonly grown in California, as typified by the Atlas varieties. They are characterized by large, long kernels weighing about 45 mg, having relatively thick hulls that tend to obscure the blue color of the aleurone when present. Variations in color of aleurone and shape of heads and kernels are found in representatives of this group.

In the common 2-rowed group, the lateral florets are sterile and greatly reduced. Only one row of kernels develops on each side of the spike. The kernels are all symmetrical and more uniform in size than those of 6-rowed varieties, but the kernels that develop at the base and the tip of the spike are often somewhat smaller than those in the center. The varieties Hannchen and Betzes, grown in the intermountain and western areas of the United States, are examples of this group. Medium-long, relatively plump kernels weighing 35 to 45 mg, with thin, white, finely wrinkled hulls and with white aleurone, are typical of the group. Except for kernel shape and size, variation in characters within important varieties of this group is less than for the 6-rowed groups.

Aberg and Wiebe (1946) gave the name *H. irregulare* to a newly defined species of barley. The median florets are fertile, but the proportions of fertile, sterile, or wanting lateral florets vary considerably. This type has been found in Abyssinia and probably originated there. The species is not represented by commercial varieties.

PRODUCTION STATISTICS

Geographical Distribution of Production

In the World.—The world production of barley has been increasing since 1950. The average estimated production for the 5-yr period 1950 through 1954 was 2.7 billion bushels. This increased to 3.3 billion bushels for 1955 through 1959, and to 3.8 billion for the next 5-yr period, 1960 through 1964. More than 4 billion bushels were produced in 1963, 1964, and 1965, the last year for which statistics are available.

Considering areas of the world, Europe produced the largest quantity of barley from 1955 through 1965 followed by Asia and North

America.   The leading barley producing countries of Europe were France, United Kingdom, Denmark, West Germany, and Spain. From 1955 to 1963, France produced more than the United Kingdom, but the reverse was true in 1964 and 1965.   The major increase in world production of barley, referred to above, occurred in Western Europe and more specifically in France and the United Kingdom.   Moderate, but less consistent increases were produced in Denmark, West Germany, and North America.   Over the last 5 yr (1961–1965) the United States produced about twice as much barley as Canada.

In the United States.—Barley production in the United States in 1967 was estimated at 373 million bushels, the lowest since 1953. As shown in Table 27, this resulted from low harvested acreage and the yield per acre was high.   Production statistics for the 20-yr period show the increased acreage and production of barley in 1954, when wheat acreage restrictions were imposed.   Acreage and production increased generally through 1959 but acreage has gradually decreased since that time.   The increase in yield per acre over the 20 yr has been gradual, but the rise for the last 5-yr period was very marked.   These were favorable years for barley, but the introduc-

TABLE 27

BARLEY ACREAGE, YIELD, AND PRODUCTION IN THE UNITED STATES, 1948–1967

	Acreage Harvested (1,000 Acres)	Yield per Acre Harvested (Bushels)	Production (1,000 Bushels)
1948	11,905	26.5	315,537
1949	9,872	24.0	237,071
1950	11,115	27.2	303,772
1951	9,424	27.3	257,213
1952	8,236	27.7	228,168
1953	8,680	28.4	246,723
1954	13,370	28.4	379,254
1955	14,523	27.8	403,065
1956	12,852	29.3	376,661
1957	14,872	29.8	442,761
1958	14,791	32.3	477,368
1959	14,918	28.3	422,383
1960	13,939	30.9	431,309
1961	12,946	30.6	395,669
1962	12,430	35.1	436,448
1963	11,566	35.1	405,577
1964	10,624	37.9	402,895
1965	9,478	43.5	411,897
1966	10,227	38.1	389,557
1967[1]	9,370	39.9	373,438
20 Year Average	11,756	31.4	366,838

[1] Preliminary estimate.

tion of higher yielding varieties and the concentration of barley production in high producing areas also contributed to higher yields.

The eight largest producing States are given in Table 28, with production figures for the 10-yr period 1958–1967. Over this period these States accounted for 75% of the barley production in the United States, and are the major producers of malting barley. The Red River Valley, occupying parts of North Dakota, Minnesota, and South Dakota, produces most of the midwestern, Manchuran 6-rowed barley. The major 2-rowed malting barley areas are in Montana, Idaho, Washington, and Oregon, and the western coast type 6-rowed barleys are grown in the central valleys of California.

TABLE 28

BARLEY PRODUCTION OF EIGHT IMPORTANT STATES FROM 1958 to 1967
MILLION BUSHELS

	1958–62	1963	1964	1965	1966	1967[1]	10-Year Average
North Dakota	83.7	104.0	91.0	105.3	88.7	82.0	90.0
California	71.4	68.0	73.3	69.3	67.3	82.7	71.7
Montana	45.2	44.7	50.5	50.7	63.6	36.1	47.1
Idaho	20.5	28.6	27.2	31.0	23.4	27.5	24.0
Washington	26.4	27.2	22.4	15.1	19.1	11.7	22.7
Minnesota	27.1	25.9	19.6	26.8	21.0	30.5	25.9
Oregon	18.1	16.7	16.3	17.0	17.8	11.9	17.0
South Dakota	11.9	8.9	5.9	8.4	7.4	12.7	10.3
US Production	432.6	405.6	402.9	411.9	389.6	373.4	414.7

[1] Preliminary estimate.

PRODUCTION METHODS IN THE UNITED STATES

Growing the Crop

For best yields of good quality grain, barley should be sown early in a well-prepared seedbed on adequate but not excessively fertile soil. In many areas of spring barley production, fall plowing of the land is recommended. This covers previous crop residues which may harbor disease organisms and permits early seeding the following spring. Barley is grown successfully on spring-plowed or disked land, but spring plowing may delay seeding date and thus increase weed competition and disease problems.

Rotation of crops is generally good practice, and barley is no exception. Where barley follows a cultivated crop, such as corn, weed competition is reduced and residues of heavy fertilization of the corn are usually adequate for good yields. However, barley following corn may be scabby unless the cornstalks are plowed under. Where barley is planted on plowed legume hay land, excessive soil nitrogen

may result in lodging.   In lower rainfall areas where summer fallowing is practiced, barley is sown usually on fallow land which supplies
more moisture and fertility.

Barley is seeded with grain drills at a rate of about $1^1/_2$ bu per
acre.   This rate may be increased or decreased somewhat depending
upon soil moisture and fertility, time of seeding, and prevalence of
weed seeds.   Where grasses or legumes are seeded with barley as a
companion crop, the seeding rate may be reduced to lessen its competition with the grass or legume seedlings.   The depth of seeding
should be from $^1/_2$ to $1^1/_2$ in.   This will depend upon soil type and
condition and surface moisture supply.

The use of fertilizers for barley is increasing.   Phosphorus and
potassium are rather commonly applied to some soils, but frequently
increased yields have resulted from nitrogen applications of 25 to
50 lb per acre drilled with the grain.   In the production of barley
for malting, heavy applications of nitrogen may increase barley
protein and reduce malting quality.   Heavy nitrogen fertilization
often results in lodging of the plants, and this interferes with normal
development of the grain.

The time of seeding varies with location.   Most winter barleys
are sown between September 15 and the last of October.   Spring
barleys are sown from April 1 to May 15, or about as early as the
land can be prepared.   In California, barleys with a spring-growth
habit are sown from late October until mid-January, and thus are
grown as a winter crop.

Harrowing of barley to control weeds is not beneficial to the crop.
When broad-leaved weeds develop early before the shade from the
grain leaves prevents their rapid growth, spraying with commercial
formulations of 2,4-D is now commonly used.   The most favorable
time to spray for weed control is when the grain is 8 to 12 in. tall.

Harvesting

Barley should be harvested only when the grain is fully ripe.   This
is especially important in the growing of barley for malting.   When
the straw is completely yellowed, the kernels feel dry to the hand and
snap when bitten, the grain moisture will usually be between 14 and
16%.   Grain harvested at 14% moisture or less will usually keep
in the bin in the cooler parts of the United States.

Combining is the cheapest method of harvesting and is used almost exclusively in large acreage production.   If ripening and drying
conditions are good, the standing grain will be combined directly.
In fields where ripening is not uniform or where many green weeds

are present, the grain is cut with a windrower and laid in swaths on a long stubble. After one or more days of drying, a pickup attachment elevates the grain to the combine for threshing. Grain in windrows deteriorates more rapidly than standing grain during rainy weather, but will be of better quality if no rain occurs.

Very small quantities of grain are cut with a binder in areas in which barley is grown in small acreages. The bundles are shocked in the field to dry and later hauled to a threshing machine for separation. This permits piling the straw in a convenient location. However, with small mobile baling equipment the straw from the combine can be picked up, baled, and stored in this manner for feed or bedding.

With either method of harvesting, the separator must be properly adjusted to thresh the grain cleanly without excessive skinning or breaking of the kernels. This is especially important for barley produced for seed or malting, since damaged kernels will not germinate properly.

Storage

At present, most barley is stored in bulk in bins of various sizes from a few hundred to many thousand bushels. For safe storage of large quantities for long periods grain should have a moisture content of less than 13% according to Tuite and Christensen (1955). Freshly harvested grain with a moisture content above 14% may heat and go out of condition. Usually moisture is not evenly distributed in harvested grain, and small areas may be wet enough to permit development of mold. This localized mold growth raises the temperature and moisture content still higher and, if it is allowed to continue, the entire lot may heat and go out of condition. Even moderate development of storage molds and increases in temperature and moisture will destroy the germination ability of barley and give it a musty odor if allowed to continue. Barley that is to be used for seed or malting requires close watching and special care in storage.

Freshly harvested grain is usually watched carefully and, if a significant rise in temperature is noted, it is moved from one bin to another which cools the grain and mixes the damp and dry portions. If very wet it should be dried artificially. Elevators for storage of malting barley are equipped with temperature-recording facilities at frequent intervals in the bins to aid in detecting the start of heating. Large growers use driers to reduce the moisture of directly combined grain to a safe storage moisture. Great care and control of conditions, especially temperature, are required to prevent destruction of germination capacity of the grain by drying.

UTILIZATION OF BARLEY IN THE UNITED STATES

Animal Feed

The major use of barley in the United States is for animal feed. Over the 10-yr period 1955 through 1964 approximately 57% of the annual production was used for this purpose, with yearly variations from 49 to 67%. The average value for the previous 10 yr was 53%, so about 50–60% normally goes into this channel. The grain is used in blends with other feed materials for all farm animals, but, at least in the past, was especially prized for bacon hogs, and sheep and lambs for show purposes.

The grain is ground and mixed with concentrates and other grains for feeding, or may be steamed and rolled to produce a flake. Dinnusson (1957) in North Dakota found that the pelleting of barley increased its palatability and efficiency as a feed for hogs. Malt sprouts, brewer's grains, and distiller's grains are barley by-products from the malting, brewing, and distilling industries respectively, and are used in mixed feeds mostly for dairy cattle.

Industrial Uses

The major industrial use of barley is for the production of malt. Estimated malt consumption for the period 1955–1964, the latest for which data is available, averaged 95.6 million bushels and ranged from 92 to 103 million bushels. Total barley production was relatively high for this period and the quantity used for malting averaged 23%. Over a longer period of time, this value would range from 25 to 30%, depending upon production and quality of the grain.

Specific varieties of three types of barley grown in definite areas of the country are usually used for malting. The 6-rowed Manchuria type, exemplified by the varieties Larker, Dickson, and Conquest, and grown in the North Central States, supplies 85–90% of the barley for malting. In years when the quantity and quality of barley from this area are low, Canadian barley may be imported to meet the requirement. From 10 to 15 million bushels of the 2-rowed varieties, Hannchen, Betzes, Piroline, and Firlbeck's III, grown in specific areas of Montana, Idaho, Washington, Oregon, and California, are required annually for malting. Relatively small quantities of the Coast-type variety, Atlas, grown in the central

California valleys are used for malting.   The Western States usually produce more good quality grain than is needed for malting.

About 85% of the malt production is used for beer manufacture, slightly less than 10% for alcohol and whiskey production, and slightly over 5% for food uses.

Human Food

The uses of barley and barley products according to Phillips and Boerner (1935) are as listed in Table 29.

The many food uses of barley and malt excluding malt beverages utilize somewhat less than ten million bushels of barley annually. An estimated three million bushels is used for production of pot and pearl barley, and small quantities of barley flour are used for baby

TABLE 29

USES OF BARLEY AND BARLEY PRODUCTS

Feed
    Livestock
    Poultry
Export
    Feed
    Malting
Pearling
    Pot barley for soups and dressings
    Pearled barley for soups and dressings
    Flour
    Feed
Milling
    Flour for baby foods and food specialties
    Grits
    Feed
Malting
    Brewers' beverages
    Brewers' grains for dairy feeds
    Brewers' yeast for animal feed, human food, and fine chemicals
    Distillers' alcohol
    Distillers' spirits
    Distillers' solubles for livestock and poultry feeds
    Distillers' grains for livestock and poultry feeds
    Specialty malts
        High dried
        Dextrin for breakfast cereals, sugar colorings, dark beers, and coffee substitutes
        Caramel for breakfast cereals, sugar colorings, dark beers, and coffee substitutes
        Black for breakfast cereals, sugar colorings, dark beers, and coffee substitutes
    Export
    Malt flour for wheat flour supplements, human and animal food products
    Malted milk concentrates for malted milk, malted milk beverages, and infant food
    Malted syrups for medicinal, textile, baking uses, and for breakfast cereals and candies
    Malt sprouts for dairy feeds, vinegar manufacture, and industrial fermentations

foods. Two-rowed varieties grown in Western areas that produce large plump kernels are preferred for pearling.

Food uses of malt are numerous, but the quantities used are small. A number of types of malt syrups for specific uses, malted-milk concentrates, enzyme supplementation of wheat flour, and breakfast foods are typical examples. As malt is used for flavor or enzyme action in most cases, relatively low concentrations are employed.

COMPOSITION OF BARLEY AND QUALITY TESTS

Structure of Barley Kernel and Methods for Identification of Varieties

A typical hulled barley kernel from the outside inward is composed of lemma and palea enclosing and cemented to the caryopsis. The rachilla lies within the crease of the kernel near the base and on the ventral or palea side. It is covered with long or short hairs. The lemma is five-nerved and somewhat angled at the nerves. The lateral and marginal nerves may have numerous or few small teeth or may be smooth. The lemma may have a depression consisting of a transverse crease at its base just above the point of attachment. The caryopsis is composed of pericarp, integuments, starchy endosperm, and germ. The outer layer of the endosperm is made up of the aleurone cells. In blue barleys, the anthocyanin pigment is blue in the alkaline aleurone cells, while the same pigment in the pericarp or hull appears as red. The aleurone of many varieties is colorless. This structure is the major enzyme synthesizing area of the kernel.

The germ is partly imbedded in the endosperm at the base of the kernel on the lemma or dorsal side and is held at an oblique angle to the axis of the kernel. The germ is composed of the embryonic axis, which develops into the seedling at germination, and the adjacent scutellum. The latter structure secretes hormones that stimulate enzyme release and synthesis. The enzymes hydrolyze constituents of the endosperm to products which nourish the growing seedling.

Threshed samples of some major varieties can be distinguished from each other with ease, while others can be distinguished only with difficulty. Two-rowed varieties with all symmetrical kernels are distinguished from 6-rowed, in which $1/3$ of the kernels are symmetrical and $2/3$ are twisted and slightly smaller. Wiebe and Reid (1961) have further separated the two types into groups on the basis of length of rachilla hairs, length of kernel, kernel color, number of teeth on lateral and marginal lemma nerves, and shape of lemma base. Additional characters are suggested for distinguishing varieties within the groups.

Varietal examples of the three major malting-barley types are described here and illustrated in Fig. 21, 22, and 23. Hannchen has covered kernels, all symmetrical (2-rowed), white, long-haired rachilla, lemma base with depression, and no to few teeth on lateral lemma nerves and few to several on marginal nerves. The hulls are finely wrinkled.

Dickson has covered kernels, $1/3$ symmetrical, $2/3$ twisted (6-rowed), short to midlong white kernels, short-haired rachilla, rough awns, few teeth on lateral and marginal lemma nerves.

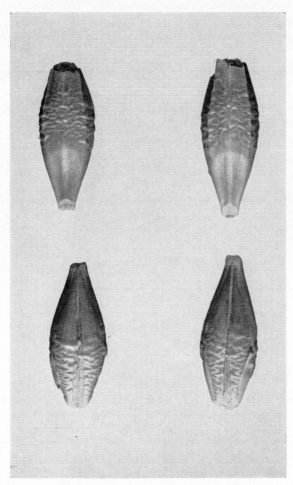

FIG. 21. KERNELS OF WESTERN TWO-ROWED BARLEY
VARIETY HANNCHEN

Atlas has covered kernels, $1/3$ symmetrical, $2/3$ twisted (6-rowed), long blue kernels, short-haired rachilla, numerous teeth on lateral and marginal lemma nerves. The three varieties described above could be easily distinguished from one another in threshed grain samples, but distinguishing Dickson from the feed varieties which fall in the same classification group is very difficult. Malting Barley Improvement Association (Anon. 1957) published a barley variety dictionary to aid in identification of commercial varieties. This loose-leaf publication is kept current for new malting varieties.

FIG. 22.   MEDIAN AND LATERAL KERNELS OF MIDWESTERN
SIX-ROWED BARLEY VARIETY KINDRED

FIG. 23. MEDIAN AND LATERAL KERNELS OF WESTERN SIX-ROWED BARLEY VARIETY ATLAS

Chemical Composition for Feed and Food Uses

The major use of barley is for animal feed. Typical proximate analyses of three barley types as given by Kneen and Dickson (1967) has been modified in Table 30. Feed barleys from drier production areas would be lower in kernel weight, higher in protein, and proportionately lower in nitrogen-free extract. Barley is used primarily as a source of carbohydrate, but the protein content is of importance for feed. In common with other cereal grains, barley is low in the

TABLE 30

TYPICAL PROXIMATE COMPOSITION OF THREE TYPES OF BARLEY[1]

Barley Type	Kernel Weight (Mg)	Hull (%)	Protein (%)	Fat (%)	Starch (%)	Fiber (%)	Ash (%)	Nitrogen-Free Extract (%)
Midwestern 6-rowed	36	12	12	2.0	58	5.7	2.7	66.6
California 6-rowed	45	14	11	2.0	58	6.6	3.0	65.4
Western 2-rowed	40	10	10	2.0	60	5.2	2.5	72.3

[1] Adapted from a section by Kneen and Dickson in Vol. 12 of the Encyclopedia of Chemical Technology, Copyright 1967 by John Wiley & Sons, New York.

essential amino acids, lysine and methionine. Comparative composition of barley, corn, oats, and wheat for 11 amino acids as tabulated by Miller (1958) are given in Table 31. Barley is higher than corn in lysine, and may be equivalent to high-lysine corns, if the Opaque -2 gene doubles the lysine content. The vitamin content of cereal grains also tabulated by Miller (1958) are given in Table 32. In general, barley is higher than corn except for pyridoxine and carotene.

TABLE 31

AMINO ACID COMPOSITION OF CEREAL GRAINS[1]

Per Cent	Barley	Corn	Oats	Wheat
Arginine	0.6	0.4	0.8	0.8
Histidine	0.3	0.2	0.2	0.3
Isoleucine	0.6	0.5	0.6	0.6
Leucine	0.9	1.2	1.0	1.0
Lysine	0.6	0.3	0.4	0.5
Phenylalanine	0.7	0.5	0.7	0.7
Threonine	0.4	0.3	0.4	0.4
Tryptophan	0.2	0.1	0.2	0.2
Valine	0.7	0.5	0.7	0.6
Methionine	0.2	0.2	0.2	0.2
Cystine	0.2	0.1	0.2	0.2

[1] Composition of Cereal Grains and Forages. Publ. 585. Natl. Academy of Sci. —Natl. Res. Council, June 1958.

TABLE 32

VITAMIN COMPOSITION OF CEREAL GRAINS[1]

Mg/Lb	Barley	Corn	Oats	Wheat
Thiamine	2.6	2.0	3.2	2.5
Riboflavin	1.0	0.6	0.8	0.6
Pantothenic acid	3.3	3.2	6.6	6.2
Niacin	29.3	11.6	8.1	28.9
Pyridoxine	▸1.5	2.6	0.6	2.4
Choline	526	274	548	424
Carotene	0.2	1.8	0.0	0.0

[1] Composition of Cereal Grains & Forages. Publ. 585, Natl. Academy of Sci. —Natl. Res. Council Washington, D.C. 1958.

In studying the nutritional value of western grown barley for young chicks, McGinnis (1958) showed increased growth rate and food efficiency when ground barley was steeped in water, dried, and incorporated into the feed. This effect was ascribed to increased carbohydrate utilization, and supplementation with crude alpha-amylase, other enzymes, and malt gave a similar response. Recent work, Burnett (1966) and Malik (1965), showed that poor growth on bright barleys was associated with high viscosity in the small intestine caused by barley gums. Water treatment and enzyme supplementation hydrolyzed the gums, reduced viscosity, and permitted improved performance. When properly treated, barley is essentially equal to corn for young chicks and turkey poults.

Physical and Chemical Requirements for Malting

Malting is the second largest use for barley and requires almost 100 million bushels annually. Significant quantities of wheat and small amounts of rye are malted in the United States, but the major malting grain is barley. Tradition may have contributed somewhat to this use, but the major reasons are related to the structure, composition, and physiology of the grain. The barley grain is somewhat firmer in structure and softens less than wheat and rye during the steeping and germination phases of malting. The attached hulls protect the plumule of the germ as it grows under the hull until the tip of the kernel is reached. In skinned kernels where the plumule is exposed it is likely to be damaged, and normal malting and modification of the kernel may be interrupted. This same situation exists in wheat and rye which thresh free from the hulls.

Good malting barleys should be plump and well filled, moderately low in protein, and capable of vigorous uniform germination. They should be free of disease, excessive staining or weathering, and high-temperature or moisture damage during storage. During the malting process, the barleys should be capable of desirable physical and chemical modification, activation and production of desired levels of several enzyme systems all with a minimum loss of dry material through respiration and rootlet development. Adequate hydrolysis of proteins and starch to permit the development of desirable flavor during kilning of the malt by the browning reaction, and without development of excessive color, is essential.

Tests for Malting Quality

The malting quality of hybrid selections or new varieties can be predicted with a reasonable degree of accuracy by determination of barley kernel weight and kernel size assortment using standard sieves and barley protein content. Barley extract is determined by using a supplementary source of enzymes to hydrolyze the barley compounds. Beta-amylase exists partly free and partly in a bound form in barley, and the latter can be freed in finely ground samples by a proteolytic enzyme such as papain or by sulfhydryl compounds. Extraction with papain is employed in the determination of diastatic power.

A more reliable evaluation of quality requires experimental malting of the barley. Rate and uniformity of water absorption in the steep and germination, as well as the percentage loss of dry matter in the process, can be obtained. The malt is analyzed for extract or total soluble constituents after mashing with water under a prescribed temperature and time schedule. Filtration of the mash gives an extract, called wort. The amount of soluble nitogen in the wort in relation to that in the malt serves as an indication of protein breakdown during malting and mashing and is a rough measure of proteolytic activity. Beta-amylase is freed during malting and can be extracted directly from the ground malt. Alpha-amylase, which is essentially nonexistent in normal barley, is synthesized during malt-

TABLE 33

TYPICAL ANALYSES OF MALTS FROM THREE TYPES OF BARLEY[1]

Analytical Item	Mid-western 6-Rowed	California 6-Rowed	Western 2-Rowed
Kernel weight, dry basis, mg	30.0	39.0	37.0
Growth, length of acrospire, inch			
$0-^1/_4$, %	1	1	0
$^1/_4-^1/_2$, %	1	5	1
$^1/_2-^3/_4$, %	4	9	4
$^3/_4-1$, %	92	85	93
Overgrown, %	2	0	2
Assortment:			
On $^7/_{64}$ screen, %	32	68	85
On $^6/_{64}$ screen, %	54	26	10
On $^5/_{64}$ screen, %	15	6	1
Through screen, %	1	0	0
Moisture, %	4.2	4.5	4.2
Extract, fine grind, dry basis, %	76.5	76.8	80.5
Extract, coarse grind, dry basis, %	74.5	74.2	79.0
Difference, %	2.0	2.6	1.5
Color, $^1/_2$ in. cell, Lovibond S. 52	1.6	1.4	1.2
Total protein, dry basis, %	12.3	11.0	10.5
Soluble protein, as % of total	40.0	35.0	38.0
Diastatic power, dry basis, °L.	135	65	100
Alpha-amylase, dry basis, 20° units	38	25	32

[1] Adapted from a section by Kneen & Dickson in Vol. 12 of the Encyclopedia of Chemical Technology. Copyright 1967 by John Wiley & Sons, New York.

ing and can be extracted and determined. Standardized procedures for the methods mentioned have been developed by the American Society of Brewing Chemists (Anon. 1958). Typical analyses of malts from the three types of malting barley as reported by Kneen and Dickson (1967) are given in Table 33.

Until more complete information is available on the chemistry of barley and the biochemistry of the malting and brewing processes, final quality evaluation of barley requires experimental malting and pilot plant brewing of the malt. The best and most widely used method for barley quality evaluation is by observations of the processing performance, analysis of resulting worts and beers, and determination of characteristics of the final beers. The malting and brewing industries are cooperating with State and Federal workers in an extensive program for the development and evaluation of improved malting barleys for the United States and Canada. Methods for barley quality evaluation were reviewed by Dickson and Burkhart (1956) and more recently by Dickson (1965).

Composition in Relation to Malting

The importance of the composition of barley and the changes taking place during malting require the development and use of specific varieties. Harris (1962A, 1962B) has reviewed in detail the chemistry of the individual groups of compounds in barley and the way in which they are degraded by the native enzymes produced during malting. One group of enzymes of barley and malt, the carbohydrases, was recently reviewed by Luchsinger (1966). In the last decade, older concepts on the parts of the kernel important in production of enzymes during germination have required modification. The production of gibberellin in the germ and its stimulation of enzyme synthesis primarily by the aleurone, has implications for future malting procedures which have been reviewed by Laufer (1964).

The four classes of proteins of Osborne, based on solubility characteristics, have been shown by modern methods to be composed of many more compounds. Immuno-electrophoretic methods indicate 17 soluble proteins in barley, many of which are changed only in molecular weight during malting. Beta-amylase has been separated into 4 fractions, differing in molecular weight, but these 4 and bound beta-amylase of barley appear to be immunologically identical (Grabar and Nummi 1967). Eighteen to 20 amino acids have been found free in barley, and the quantity of all increases during malting.

Barley starch is similar to that from other cereals, except for granule size and shape. Recent studies indicate that the amylopectin fraction is broken down during malting more extensively than amylose. The quantity of barley gums, and enzyme systems that attack them, are important in the physical modification of barley during malting and the degree of extraction during subsequent processing. Total starch is related to protein content and, after malting, to malt extract yield.

Beta-amylase is the only enzyme of importance to malting and malt processing which is abundant in normal mature barley. The others are synthesized during malting, probably mostly in the aleurone cells.

BIBLIOGRAPHY

ABERG, E., and WIEBE, G. A. 1946. Classification of barley varieties grown in the United States and Canada in 1945. US Dept. Agr. Tech. Bull. *907*.

ANON. 1957. Barley Variety Dictionary. Malting Barley Improvement Association, Milwaukee, Wisc.

ANON. 1958. Methods of Analysis, 6th Edition. American Society of Brewing Chemists, Madison, Wisc.

BURNETT, G. A. 1966. Viscosity as the probable factor involved in the improvement of certain barleys for chickens by enzyme supplementation. Brit. Poultry Sci. 7, 55–75.

CLARK, H. H. 1967. The origin and early history of cultivated barleys. A botanical and archaeological synthesis. Agr. History Rev. *15*, 1–18.

DICKSON, A. D. 1965. Barley and malt: Fifty years of progress in cereal chemistry. Cereal Sci. Today *10*, 284–290.

DICKSON, A. D., and BURKHART, B. A. 1956. Evaluation of barley varieties for malting quality. Proc. Am. Soc. Brewing Chemists, 143–155.

DINNUSSON, W. F. 1957. Pelleting rations for swine. N. Dakota Agr. Expt. Sta. Bimonthly Bull. *20*, No. 1, 13–16.

GRABAR, P., and NUMMI, M. 1967. Recent immuno-electrophoretic studies on soluble proteins in their transformation from barley to beer. Brewers Dig. *42*, No. 6, 68–73.

HARRIS, G. 1962A. The enzyme content and enzymic transformation of malt. *In* Barley and Malt. A. H. COOK (Editor). Academic Press, New York.

HARRIS, G. 1962B. The structural chemistry of barley and malt. *In* Barley and Malt. A. H. COOK (Editor). Academic Press, New York.

KNEEN, E., and DICKSON, A. D. 1967. Malts and Malting. *In* Encyclopedia of Chemical Technology. John Wiley & Sons, New York.

LAUFER, S. 1964. New concept regarding mechanism of malt modification. Am. Brewer *97*, No. 5, 75–77.

LUCHSINGER, W. 1966. Carbohydrases of barley and malt. Cereal Sci. Today *11*, 69–75, 82–83.

MALIK, M. Y. 1965. Studies on the nature of chick growth-inhibiting factor in barley. West Pakistan J. Agr. Res. *3*, 154–161.

McGINNIS, J. 1958. Enzyme supplements for grains—some implications. Foodstuffs *30*, No. 20, 58.

MILLER, D. F.   1958.   Composition of Cereal Grains and Forages.   Publ.
   585.   Natl. Acad. Sci.—Natl. Res. Council, Washington, D.C.
NUTTONSON, M. Y.   1957.   Barley—Climate Relationships and the Use
   of Phenology in Ascertaining the Thermal and Photo-thermal Require-
   ments of Barley.   Am. Inst. Crop Ecology, Washington, D.C.
PHILLIPS, C. L., and BOERNER, E. G.   1935.   Barley and barley malt.
   US Dept. Agr. Bur. Agr. Econ.
SHANDS, H. L., and DICKSON, A. D.   1953.   Barley—botany, production,
   harvesting, processing, utilization and economics.   Econ. Botany 7,
   3–26.
SPECTOR, W. S.   1956.   Handbook of Biological Data.   W. B. Saunders
   Co., Philadelphia, Pa.
TUITE, J. F., and CHRISTENSEN, C. M.   1955.   Grain storage studies.
   XVI.   Influence of storage conditions upon the fungus flora of barley
   seed.   Cereal Chem. 32, 1–11.
WEAVER, J. C.   1950.   American Barley Production.   Burgess Publishing
   Co., Minneapolis, Minn.
WIEBE, G. A., and REID, D. A.   1961.   Classification of barley varieties
   grown in the United States and Canada in 1958.   US Dept. Agr. Bull.
   1224.

CHAPTER 5

H. L. Shands | # Rye

HISTORY OF RYE CULTIVATION

Rye is among the less important cereal crops of the Americas, but it is of major importance in Europe and parts of Asia where it is used as a bread grain. The exact time that rye was brought under cultivation is not known. It is thought to be one of the crops more recently domesticated. Whether it is more recent than the hexaploid bread wheats is a matter of speculation. Rye seeds have not been found in the excavations of the cliff dwellers or early Egyptians nor was the crop pictured on early coins as was barley and emmer. Vavilov (1926) thought that cultivated rye arose from the rye-weed originally growing in crops of barley and wheat. Cultivation of the crop probably started before the Christian era.

As cited by Kranz (1957), Zhukovsky and Schiemann suggested that cultivated rye may have arisen from the two independent sources, *Secale ancestrale* Zhuk. and the wild types of *Secale montanum* Guss. Kuckuck and Kranz (1957) after studying Iranian rye populations, suggested that cultivated rye may have originated from several sources as hypothesized by Roshevitz (1947). He indicated that perennial wild ryes could have given rise to annual wild ryes which in turn could have been the forerunners of cultivated rye. Kranz (1963) later stated that Iranian ryes were transitional between wild ryes and cultivated ryes.

Vavilov stated that rye was undoubtedly introduced into cultivation simultaneously and independently at many localities. He found greatest botanical diversity in Afghanistan, Persia, Transcaucasia, Asia Minor, and Turkestan, finding as many as 18 botanical varieties in Afghanistan. Zhukovsky (cited by Vavilov) found 14 varieties in Georgia and Armenia. The weed-rye of Afghanistan, Persia, and Turkestan had types with closely investing lemmas, nonshattering spikes, and had adpressed awns. The weed-ryes of Armenia and Asia Minor had spreading awns with noninvested kernels that resembled cultivated ryes.

The cultivated winter rye may have entered Europe from two sources according to Vavilov. One was from Transcaucasia and

---

H. L. SHANDS is Professor of Agronomy of the University of Wisconsin.

the other from Turkestan, Afghanistan, and adjoining regions. Then it probably spread northward during times when cultural practices were unrefined and probably was often sown in preference to wheat because rye could withstand more adversities. Khush (1963) thought that rye was domesticated at several places independently and at different times.

Adversity is no stranger to rye even to the present day. Vavilov stated that the peoples of Persia and Afghanistan considered rye to be a noxious weed and found it difficult to control. One reason it was considered noxious was that the seeds were difficult to separate from those of wheat. As late as 1925 the Board of Agriculture of northern Caucasus recommended mowing down rye while in bloom, but cautioned not to injure wheat. Farmers who adopted this practice in 1925 were freed of taxes. Such an inducement in the western world today might have far-reaching effects in reducing this crop.

Because of its competitive ability among other plants, rye may have increased at the expense of wheat acreage in northern Europe for a period of several centuries until better adapted wheats or better cultural practices were found for wheat. For more than 100 yr rye has been losing acreage in European countries and the end may not yet be in sight. Jasny (1940) pointed out that the use of rye for bread permitted many economies in baking, merchandising, etc.; yet the taste for and desire for rye bread was diminishing in heavy rye-producing European countries. Ljung (1948) indicated that Sweden grew rye and wheat in a ratio of 4 to 1 at the beginning of the twentieth century, but by 1948 wheat production was about equal to rye in that country. By 1966 the ratio fell to about 1 to 6. The trend has been downward in the western hemisphere. Yet there are vast areas in the Old World where rye is certainly the most dependable grain crop and undoubtedly it will remain popular.

Rye was introduced into northeastern United States during colonial times by English and Dutch settlers according to Collier (1949). From this area rye spread southward and westward where acreage rose for a period of years, but more lately receded.

CHRONOLOGICAL DATA OF RYE PRODUCTION

World Production

Total world production of rye for the past 25 yr has averaged near 1.4 billion bushels according to Agricultural Statistics of the US Dept of Agr. Since the legal weight for rye is 56 lb per bushel, rye production may be compared easily with that of corn and wheat, but there needs to be an upward adjustment of bushels if compared with rice, barley, or oats.

For the 5-yr period 1960–64, the average world production was almost 1.3 billion bushels. Previous to that time, there was moderate fluctuation notably during, and immediately after, World War II. The average production from 1925 to 1940 was somewhat more than 1.7 billion bushels. In general, total production of rye has remained fairly constant but with a slight downward trend during the past ten years.

Russia produces the greatest amount of rye annually—in fact, this country produces about half of the total world output. However, rye is rather intensively cultivated in Poland, West Germany, East Germany, and Czechoslovakia. It is interesting to note that there is more rye than wheat acreage in both East Germany and Poland. The United States and Canada produce only a small percentage of the world supply.

Other countries producing a significant amount of rye are France, Spain, Austria, Netherlands, Hungary, and Argentina. Turkey produces as much as the United States. Argentina's production of rye has fluctuated pronouncedly, ranging from 3.2 million bushels in 1951 to more than 52 million bushels in 1952 and 12 million bushels in 1966.

Production in the United States

Production of rye in the United States has varied over a wide range since beginning records in 1869. Production in 1869 was near 16 million bushels; it gradually rose to 29 million bushels in 1909, increased rapidly until 1919, and then reached a high point of more than 101 million bushels in 1922. This bulge in production may have been partly accounted for by several rust years for wheat during World War I. The average production in the United States for the 5-yr period 1939-43 was 40 million bushels and for the 5-yr period from 1960-64 was 33 million bushels. This indicates a downward trend in interest in rye in the United States.

The North Central states produce the greatest quantity of rye within the United States. During the period of 1961-65 there were 6 states in this area that produced more than a million bushels for the 10-yr average. North Dakota produced nearly 10 million bushels while South Dakota, Nebraska, Kansas, Minnesota, and Washington followed with decreasing production. In 1967, Washington's harvest dropped to $^1/_3$ of a million bushels while Georgia produced more than a million in that year.   Production has fluctuated partly in response to weather conditions, wheat acreage allotments, and prices received.

Acre Yields

Acre yields of rye reach a low point in the Union of South Africa where 5 to 7 bushels per acre is the average.   In contrast to this, Belgium, Denmark, West Germany, Netherlands, Switzerland, and the United Kingdom averaged more than 40 bushels per acre in the 1960-64 period, according to Agricultural Statistics of the US Dept. of Agr.   Wheat outyields rye by small to large margins in these countries except Switzerland.

BOTANICAL CLASSIFICATION OF RYE

The genus *Secale* belongs to the tribe *Hordeae* of the grass family. Rye is almost completely cross pollinated, although autogamous forms have been noted.   The inflorescence of rye is a spike or "ear" that resembles that of wheat.   The spikelets are arranged alternately and are placed flatwise against a zigzag rachis.   Most forms of *Secale* have spikes that taper toward the tip.   Spikelets are usually two-flowered except under thin planting or fertile soil conditions. They are composed of 2 thin, narrow glumes that subtend the 2 florets each of which is partly enclosed by a narrow lemma and palea. Each floret contains 3 stamens and a pistil which, after fertilization, develops into a 1-seeded fruit known as the caryopsis and is called the kernel or grain.   All domesticated rye grain threshes free.

Key to *Secale* Species

Antropov and Antropov (1948) presented a synopsis of species and botanical varieties of the genus *Secale* including the key which follows (Flaksberger *et al.* 1939).

These workers classified 9 species as having various degrees of rachis shattering and 1 having a "tough" or nondisarticulating spike, the latter being *Secale cereale* L.   Of the 10 species, 6 were perennial, 4 annual.   Though *S. anatolicum* is listed as a perennial, nursery cultures of this species at Madison, Wis., have responded as a regular winter annual.   *S. cereale* was described as having winter or spring annual forms.   They described 28 botanical varieties of *S. cereale* that do not disarticulate and 18 that shatter.   They also described 94 subvarieties within the *"vulgare"* grouping of Koernicke.   Their basis for separating these species, varieties, and subvarieties were glume shape, awn type, spike density, lemma pubescence, color of

FIG. 24.   RYE, *Secale cereale* L., IN FLOWER

Note exposed stamens after shedding pollen.

lemma, ligule characteristics, and auricle color.   In regard to species grouping Kranz (1957) pointed out that Schiemann divided *Secale* into two groups; I., the wild types having the following species: S. *sylvestre*, S. *montanum*, and S. *africanum*, and II., those being cultivated: S. *fragile* and S. *cereale*.   Roshevitz (1947) divided the *Secale* genus into *Silvestria*, *Kuprijanovia*, and *Cerealia* groups.   He recognized all of the species listed in the key below in addition to S. *daralagesi* Thum., S. *afghanicum* (Vav.) Roshev., S. *dighoricum* (Vav.) Roshev., and S. *segetale* (Zhuk.) Roshev.   In his monograph on rye, Deodikar (1963) listed four additional species that he said are fairly valid.   They are *S. aestivum* Upensky, *S. arundinaceum* Trantz., *S. sibericum* Zhuk., and *S. turkestanicum* Bensin.   Another worker has recognized the binomial *S. chaldicum*.   Nakajima (1962) found much regularity in meiosis of several interspecific hybrids of *Secale* indicating close relationships.   Khush (1963) crossed primitive ryes with Merced, *Secale cereale*, and found similarity of genomes.   His work substantiated the findings of Nakajima (1962).   Khush thought that all rye species could be recognized as *Secale cereale* with subspecies. This is a valid viewpoint, but if adopted would necessitate trinomials and accompanying difficulties.

Cultivated Varieties Grown in the United States

Cultivated agricultural varieties are less numerous than in other small grain crops.   These varieties are plastic and variable partly because of cross pollination and different growing conditions.   Hancock and Overton (1960) observed that Balbo seed derived from certain areas of the United States produced plants much more prostrate in winter than plants from Tennessee-grown seed.

The Abruzzi variety of Italian origin has been grown widely in the southern states for a long time.   Wesser, described by Morey (1964), and Explorer, described by Wells (1966), are newer varieties for southern states.   Elbon and Gator are also used in this area.   Balbo was introduced into the United States in 1932 according to Mooers (1933); it has been popular in the central tier of states.   Rosen has been grown even longer in this area and Michigan.   Spragg and Nicolson (1917) reported on the origin of Rosen.

The Dakold variety has long been considered winterhardy.   It was distributed by the North Dakota Agricultural Experiment Station in 1922.   Pierre is grown in South Dakota.   Imperial with light amber kernels was once known as "white" rye.   It was replaced by Adams which was distributed by the Wisconsin Agricultural Experiment Station in 1953.   Caribou, distributed in the same year in Minnesota, gave promise of good yields according to Robinson and Koo (1954).   This variety is a sister of Antelope described by Harrington (1953).   Petkus, Tetra Petkus, and "Von Lochow" of European origin, are grown to a limited extent in Canada and the United States. McBean (1966) listed Dakold and Petkus as parents of Frontier, a new variety that might be useful in Canada.   The performance of several important rye varieties tested under Indiana conditions was described by Caldwell *et al.* (1958).

Key for species determination[1]

I. Spike breakable, disarticulates at maturity
  A. Glumes with awns 1.5–2.3 cm long, 2 to 3 times as
    long as the glumes. Annual plants, short   *S. silvestre* Host
  B. Glumes without awns or with short awns (less than
    1.5 cm)
    1. The rachis disarticulates down to the base.
      Caryopsis laterally compressed. Plants very
      robust, stems up to 3 meters tall, annual   *S. ancestrale* Zhuk.
    2. At maturity upper ¹/₃ to ²/₃ of the rachis dis-
      articulates
      a. Annual plants, short and wild   *S. Vavilovii* Grossh.
      b. Perennial plants, 60–150 cm in height
        X. Plants entirely covered with soft pub-
          escence   *S. ciliatoglume*
              (Boiss) Grossh.

        XX. Plants glabrous or with pubescence on
          outer epidermis of lower leaves
          O. Spikelets small, lemma 9–11 mm in
            length
            o. Glumes unequal; lemmas with
              very fine and short hairs on
              the keel   *S. africanum* Stapf
                     and Hook.

            oo. Glumes equal; lemma always
              with short hairs   *S. anatolicum* Boiss.
          OO. Large spikelets, lemma 14–18 mm
            long
            o. Stem 150 cm tall and 5–6 mm
              in diameter. Leaf 1–2 cm
              wide. Ligule 3–4 mm long   *S. Kuprijanovii*
            oo. Stem up to 90 cm in length and
              2–3 mm in diameter. Leaf
              up to 5 cm wide with ligule
              up to 2 mm in length
               ξ. Plants grey-green   *S. montanum* Guss.
               ξξ. Plants with small blue spots   *S. dalmaticum* Vis.
II. Spikes tenacious (tough) (except the variety *afghani-
    cum* Vav., whose spike disarticulates down to the
    base). Glumes with awns, very seldom without awns.
    Lemma with awns shorter than the spike. Caryopsis
    enlarged. Plants spring or winter annuals   *S. cereale* L.

Only a small amount of spring rye is grown in North America.
Merced is sown in California and Prolific in other areas.

Varieties Grown in Europe and Russia

One of the most outstanding varieties grown in Europe is Petkus
selected out of Pirnaer and Probsteier, Landsort varieties by Dr. F.
von Lochow. The first selection was made in 1881, with first distribution in 1889. From this variety many selections have been
made. There is a Petkus spring rye which yields well under favorable conditions, but usually less than winter Petkus.

---

[1] Apparently original in Russian; translated into Spanish by José Buckevicrus
with resumé by Castulo Cialzeta. Translation of Spanish into English by
Eduardo Neale-Silva and H. L. Shands.

Other varieties grown in western Europe are von Rumker, Dominant, and Heertvelden, the latter two being distributed in 1953. According to Sneep *et al.* (1967) Dominant was bred from a cross between von Lochow's Short Straw and Brandt's Marion. Heertvelden, from the cross between Ottersume and Petkus, is resistant to eelworm (nematode). Some of the varieties grown in Northwestern Europe are Star, Steel, King's II, Sangaste, Borris, Ensi, Varne, and Zelda.

Nuttonson (1958) studying climate relationships of rye in North America, Poland, Czechoslovakia, and the Soviet Union stated that the most widely distributed varieties in Russia were Vyatka, Vyatka 2, Vyatka Moskovskaya, Kazanskaya, and Lisitsina. Vyatka 2, Zazerskaya, and Zima were reported to be in great demand in Russia. He also described twelve of the the most widely grown varieties in Russia.

The Vyatka variety is grown in all regions of northern and the northeastern parts of the Soviet Union. It is also grown widely in the nonChernozem belt and in many areas of Siberia where snow persists during the winter. More recently Improved Vyatka (Vjatka), Jygeva, and Kolarovka are said to be still better varieties in Russia.

Improvement Problems and Accomplishments

In his rye monograph containing almost a thousand references, Deodikar (1963) laid considerable emphasis on various phases of improvement. His chapters contain topics on objectives, intraspecific crosses, interspecific crosses, intergeneric crosses, polyploidy and mutation, and a summary of the present position. A complete review of breeding investigations is beyond the scope of this chapter. It is well to point out that much interest and time has been given to rye improvement in Western Europe and especially in Sweden where Heribet-Nilsson (1916) described inbreeding studies and cited Rimpau's 1877 paper. Arne Müntzing of Sweden and his numerous coworkers have explored various facets of rye investigation.

Rye breeding programs have as common objectives the development of rye that will be attractive for the grower, processor, and user. Environmental factors including soil types play an important part in the response of the plant. Grazing properties are important where rye is used for forage or pasture; otherwise grain quantity and quality are paramount. To achieve economical production of grain, plant breeders seek high test weight per bushel (a high weight-to-volume ratio), lodging resistance, and shorter strawed types with a high ratio of grain-to-straw. A high milling (flour extraction) quality is also desired which in turn may be achieved by lighter colored aleurones. Most of these objectives are enhanced by disease resistance, winterhardiness, and wide adaptation. Finally, nutritional aspects must ever be kept in mind to help feed many whose diets are low in quantity or quality according to some standards.

Rye improvement methods have varied because no well established method has been accepted widely. Nearly every variety was produced by a slightly different breeding method. Mass selection in one form or another was the basis for originating varieties several decades ago. Varieties depend somewhat on the breeder's concept as to the type to which they should conform. Close breeding was used to establish some varieties while others have been derived by compositing residue seed lots of plants that were progeny-tested after close selection. Others have been purified by removing undesirable portions of the population. The Pierre variety was developed by compositing 16 inbred lines isolated from Dakold and Swedish rye varieties according to Grafius (1951). Sprague (1938) used similar procedures in developing Raritan. Inbreeding to isolate superior lines has been attempted by Russian workers as well as those of United States and Canada. Cyclic inbreeding after hybridizing inbred lines has resulted in superior inbred lines; but none has equalled the open pollinated sorts for yield and vigor.

Self-sterility has been widely observed though inbred lines with more than 50% self-fertility are now available. Kranz (1957) indicated than an Iranian rye flowered cleistogamously and was moderately self-fertile. Though Leith and Shands (1938) stated that properly chosen inbred lines gave hybrid vigor after being crossed, no method of utilizing this principle has been found. Pivnenko (1963) reported that heterosis was responsible for an increase of 420 lb per hectare or about 3 bu per acre more than the higher yielding parent. Warren and Hayes (1950) studied rye polycross progenies for yield but did not develop a variety after using this breeding method. Ferwerda (1956) proposed a system of breeding rye whereby two varieties are used for interpollination after clones have been tested and found superior on the basis of progeny tests. Clones found desirable after testing in intravarietal and intervarietal crosses would be increased in isolation and then equal quantities of seed mixed for growing commercial seed. After further study, Ferwerda (1962) reported that one cycle of recurrent selection gave only limited progress. The effectiveness of recurrent selection largely depended on finding clones with good combining ability and the detection of out-

standing combiners.   Popov (1962) proposed a plan of five years that
he called vegetative halves for breeding winter rye.   Individual plants
are selected, cloned, and then $^1/_2$ of each desirable clone is retarded
by a short day regime.   On the basis of progeny tests the delayed
halves could then be controlled for hybridization.   Further progeny
tests would be needed.

FIG. 25.   SPIKES AND KERNELS OF ADAMS AND TETRA PETKUS RYE

Upper left, two spikes of Adams rye.   Lower left, enlarged
kernels of the Adams variety.   Upper right, two spikes of Tetra
Petkus. Lower right, enlarged wrinkled kernels of Tetra Petkus.
These kernels show incipient sprouting.

Tetraploids.—Ross (1953) gave a history of Tetra Petkus rye which was obtained by doubling the chromosome number by the use of colchicine. This variety has stiff straw, large kernels, broad leaves, and delayed heading and ripening. It is grown commercially to a limited extent in Canada and the United States. Brown and Nelson (1957) at Michigan could find little advantage of this variety over diploid Rosen. These workers, as well as Koo (1958), and others, suggested that diploid rye should be isolated from tetraploid rye if good seed set were to be expected in Tetra Petkus. Patterson and Mulvey (1954) comparatively grew Tetra Petkus and von Rumker, a diploid type, and did not find severe seed reduction because of pollen interference. Müntzing (1954) hybridized 3 tetraploid forms and found 18–19% yield increase over that of the parents. He used spaced plants for the tests.

Because Tetra Petkus was a significant departure in agronomic type from the ordinary diploids, there was increased interest in breeding by means of tetraploidy. Workers in Poland, Finland, Norway, and Sweden gave reports favorable to tetraploids over their diploid counterparts. Tetraploids were said to respond more favorably to increased nitrogen fertilization giving more green fodder and crude protein according to Banneick's (1962) tests in Germany. Increased grain yield, protein, and vitamin content were found in Poland. However, drying of late maturing tetraploids at harvest became a problem in Finland. Most workers found larger seeds, more sterile florets, and more ergot. Compared with diploids, Ruebenbauer and Biskupski (1963) reported that tetraploids had higher protein, lower weight-to-volume ratios, but better baking quality with larger loaf volume. Fröst and Ellerström (1965) found a strong negative correlation between seed set and protein content of seeds. Torop and Pahmova (1966) thought that tetraploids could be selected effectively for higher per cent seed set after hybridization. Lundquist (1966) gave a detailed report on heterosis and inbreeding depression in autotetraploid rye. Silván (1967) assumed that application of polyploidy concurrently with inbreeding would be a successful means of rye improvement. Nine of his 23 tetraploids were more productive than the diploid types. Hilpert (1957) stated that tetraploids were of great interest to breeders, but that no tetra rye had been important in practical farming. In personal communications in 1967, several European plant breeders indicated a continuing interest in tetraploids even though more than 97% of European rye acreage was still of the diploid type.

Plant character differentiation has been one of the by-products of the inbreeding portions of larger improvement programs. Münt-zing (1963) noted continuing segregation for chlorophyll deficient types after 30 yr of selfing. Sybenga and Prakken (1963), Dumon and Laermans (1963), and Shands and Forsberg (1964) have de-scribed a moderate number of plant characters, most of which were in inbred lines. If these studies were pursued vigorously, promising results of inheritance and chromosome mapping might be forthcom-ing. Fröst (1966) observed significant differences in phenolic com-pounds of inbred lines.

Accessory or B chromosomes have been the subject of many stud-ies by Swedish as well as other workers. Various chromosome frag-ments and their behavior have been noted. Fröst (1963) found a positive correlation between protein content and frequency of B chro-mosomes, and thought that accessories might stimulate protein pro-duction. Higher kernel weight and higher protein combination may have resulted from fewer kernels. Müntzing (1962) noted that higher seed set percentages accompanied lower protein and lower accessory number. Deodikar (1963) concluded that inherent limitations in rye improvement have resulted in self-imposed evolutionary stag-nation at the diploid level.

Rye-wheat hybrids have been attempted for several scores of years, but Deodikar (1963) stated that rye still shows resistance to participate in a cartel union.

Segments of a rye chromosome have been added to a wheat chro-mosome as described by Driscoll and Anderson (1967) to improve leaf rust and mildew resistance in wheat.

Amphidiploids between *Triticum* and *Secale* giving 2n chromosome number forms of 56 and 42 have been called *Triticales*. Müntzing (1939) and more recently Sanchez-Monge (1956), Sadanaga (1956), and still others have utilized different *Triticum* species and *Secale cereale* for compounding different *Triticales*.

The Canadians have renewed efforts to produce 42 chromosome *Triticales*, and McGinnis[2] expressed restrained optimism that super-ior *Triticale* selections would be forthcoming. Unrau and Jenkins (1964) showed that hexaploid *Triticales* flours were not satisfactory for milling and baking, but could be blended with flour of hard red spring wheat Pembina and still give loaf volumes as goood as those when pure Pembina flour was used. According to McGinnis pre-liminary feeding trials have been erratic concerning total feed con-sumption and gains by livestock.

---

[2] Personal communication 1967.

GEOGRAPHICAL DISTRIBUTION OF PRODUCTION AND CONSUMPTION

Patterns of Distribution

Rye is one of the most widely grown small grains. It can be produced as far north or south as other cereals, but finds little use in tropical areas. The Americas are low producers of rye, yet the crop will grow and produce well under a wide range of conditions. Other higher income crops are chosen in preference to rye except where rye has a particular advantage in being able to produce under very low fertility or other adverse conditions.

There is a tendency to grow rye for forage purposes in warmer locations and for grain in colder climates. Nuttonson (1958) stated that winter rye production reaches beyond the Arctic Circle, or within a short distance of the northern limit of spring barley.

In comparing rye and wheat production in the Americas, the ratio is more than 50:1 in favor of wheat, while in Europe the ratio is about 4:1. Very significant changes in the production of these 2 crops have taken place in Russia since 1950 at which time the acreages were near a ratio of 9 or 10:7. In 1966 the ratio changed to 4:1 with wheat production being greater. Wheat has gained about 70 million acres and rye lost 30 million since 1950. Even so, rye is still grown extensively in Russia, the reason being that rye can withstand lower temperatures than wheat, with or without snow cover and still grow where moisture and soil fertility are low. The root system of rye is vigorous and weighs about half again more per acre than wheat, according to Nuttonson's (1958) report of Russian data.

Volin (1951) stated that rye acreages declined in Russia from 1925 to 1935. Wheat is given preference on black soil areas and rye on the nonblack soils which are generally inferior. Yields are near 14 bu per acre in Russia compared to about 21 in the United States.

Uses

The domestic uses of rye on US farms for 1956-65 are given in the 1965 US Argicultural Statistics. Seed usage was almost constant while that fed to livestock varied more, and the amount sold varied still more. Total exports ranged from 7,000 bu in 1953 to more than 20,000,000 in 1962, while imports ranged from about $1/2$ million to 13.5 million bushels.

In Europe the rye grain is used largely for bread-making purposes. Undoubtedly there are numerous other uses, such as flat bread, porridge, and alcoholic products.

Jasny (1940) indicated that heavy consumption of rye bread was centered in Germany, Austria, Scandinavia, central and northern Russia as well as in Russian-dominated Baltic states, Czechoslovakia, and Poland. Mixing wheat and rye flour has increased. The use of pure rye bread has tended to be confined to the poorer populations which may be related to the belief that rye bread staves off hunger pangs longer than other breads. This coincides with the German saying that rye bread keeps one satisfied longer.

This may be the result of lesser palatability of rye and the longer time required for digestion. In using rye for flour, it is usually diluted with wheat flour, in order to make a more palatable slice of bread. However, a few bakeries in the United States use only rye flour for making bread in areas where immigrant populations are high.

Rye is used to a limited extent in feeding various classes of livestock. Delwiche *et al.* (1940) gave several formulas for using rye as a concentrate in feeding dairy cattle, fattening cattle, swine, brood sows, and poultry. Below are given typical formulas for dairy mixtures and for pigs fed in dry lot.

Dairy Mixture to Be Fed with Mixed Grass and Legume Hay, and Corn Silage or Roots; About 17% Protein

	Lb
Ground rye	400
Ground oats, barley, or corn	250
Wheat bran	200
High protein concentrates[3]	150
Salt, preferably iodized	10
Total	1,010

For Pigs in Dry-Lot

Ground rye	30.0
Ground corn or barley	59.5
Linseed meal or soybean oil meal	5.0
Ground alfalfa hay	5.0
Salt	0.5

It can be seen that about 40% of the dairy mixture is composed of ground rye, while 30% is used for the pig ration. Formulas were also given for poultry use. However, the amount of rye was only about 20% in this case.

---

[3] High protein concentrates may be linseed meal, cottonseed meal, gluten meal, soybean meal, or the equivalent amount of protein by way of some other protein concentrates.

The rye crop is used quite widely for pasturage in the United States. To a lesser extent it is used for hay and cover cropping. Since rye is very winter-hardy, adapted types produce green feed during the winter in much of the southern half of the United States. Rye is used also in early fall and early spring pasturing in other sections of the country.

While there have been complaints about rye flavoring the milk from dairy cows, part of the trouble may be caused by distasteful weeds such as wild onions growing in the rye. There is no convincing proof that the rye pasture flavor is any more undesirable than that of other pasture plants. Furthermore, the pasture flavor of the milk may be reduced by removing the cows from the pasture 2 to 4 hr before milking.

GROWING, HARVESTING, AND STORING RYE

Culture

Culture of rye has been reviewed by Martin and Smith (1923), Leighty (1916), and Delwiche *et al.* (1940). These and other writers pointed out that rye will grow on good soils, but for economic reasons is sown on soils that are low in fertility and moisture holding capacity. Although rotation is desirable, rye can be grown for several years on the same fields without the benefit of rotation as is often done in Europe. Rye may follow corn in the rotation in the central states. Corn most often follows a hay crop that was first established by use of a spring grain crop. The rotation would then be rye, spring grain used as a companion crop, hay, and corn. Rye fits into numerous other rotations, but often is omitted intentionally in areas where it volunteers in the winter wheat crop. Sometimes rye is pastured in the fall and in the spring, then plowed under and followed by corn. Rye has been used occasionally as a companion crop for establishing grass and legume seedings that were spring seeded on frozen ground. Delwiche *et al.* (1940) listed several rotations for various soil types in Wisconsin. Corn, potatoes, or sugar beets might precede rye. The crop might be followed by spring grain seeded with grasses and legumes which can be used for hay or pasture for two or more years. Rye also might be substituted for wheat in rotations.

In Russia, Williams (1952) recommended that wheat and rye follow grass in a grass-arable rotation. Sugar beets or potatoes could follow rye. This was desirable because beets and potatoes did well where soil nitrogen was relatively low. A rotation of rye, sugar beets, and rye is followed in western Europe.   There eelworm is enough of a problem to enforce rotation.   Volin (1951) discussed rotations followed in cropping systems in Russia where rye makes up a considerable percentage of the acreage.   He stated that Williams had recommended that grass-legume crops be used in a soil-improving system.   The results were not immediately successful because of the shortage of grass seed.   It must be kept in mind that, as pointed out by Nuttonson (1958), a great deal of Russian farmland receives low annual rainfall, and moisture conservation must be practiced. Therefore, the fallow system is part of their rotational plan.

Nutrient Requirements

Rye utilizes the major and minor elements in producing a crop. Nitrogen, phosphorus, and potassium are the major elements usually applied.   Lime for the provision of calcium for soil acidity correction and nutrition is applied especially for the legume in the rotation. In many rye soils no fertilizer is applied at planting time, but instead the crop is topdressed with a nitrogenous fertilizer.   This is especially true where rye is sown as a cover and grazing crop between cotton rows in the South.   Nevertheless, Delwiche *et al.* (1940) and Albert (1951) recommended complete fertilizer for increased grain yields.

Delwiche *et al.* (1940) pointed out that applications of phosphate and potash could be made at seeding time using rates of 200 to 400 lb per acre.   They indicated that nitrogen could be used in the fertilizer applied at seeding and that formulas in the neighborhood of 10–14% N, 6–8% P_2O_5, and 9–12% K_2O were adequate.   They also suggested that 20 to 60 lb of elemental nitrogen could be applied as top dressing in the spring after growth starts.   For each 15 to 20 lb of nitrogen they expected an increase of 5 bu of grain per acre.   Albert (1951) stated that when soil phosphate and potash were low in availability, 500 lb of 12-4-8 or an equal amount of 10-10-10 may be applied at seeding time.   He also indicated that 5–7 bu or more of rye grain and 500 lb of straw would be produced per acre with each 20 lb of nitrogen used as top dressing.   Widdifield (1953) in North Dakota suggested that 50 lb of phosphate and fertilizer such as 0-43-0 or its equivalent could be used for rye on fallow ground.   On nonfallow ground he suggested the use of 70 lb per acre of a fertilizer such as 8-32-0.

Soil Management

Rye responds moderately well to carefully managed soil but is sown on sandy soil or on soils of low fertility.   In a 20-yr soil management test at Brookings, South Dakota, Puhr (1962) noted that rye used the least soil nitrogen and most phosphorus when compared to wheat, oats, barley, and corn under continuous cropping. Straw was not returned to the soil, but stubble and corn stalks were returned. Puhr stated that additions of 20–30 lb of nitrogen and 20 lb of P_2O_5 were enough for maximum yields under the conditions of his experiment.

Seedbed Preparation

Soil preparation is usually kept to a minimum. When rye is sown on summer-fallowed land the crop will act as a competitor for weeds and check their growth according to Martin and Smith (1923). Where rye follows a corn crop that was ensiled, it is desirable to plow the land. However, rye land may be prepared by disking only. Rye is sometimes drilled into grain stubble without previous preparation. Drill rows are 7 or 8 in. apart.

Variety Choice

The choice of a variety depends somewhat on whether the crop is to be used as a cover crop, pasturage, or for grain. If the purpose is for grain production, the variety must be winter hardy enough to withstand winter conditions in the area grown and should be of a type preferred by the cash market. Some buyers of rye grain prefer kernels of varying colors while others prefer rye that has a clear amber color resembling that of wheat. The germination of the rye seed should be tested before sowing because rye loses germination capacity more rapidly than do other small grains. Farmers generally increase seeding rates when rye has been stored more than a year. Seed may be treated with an organic volatile mercury compound if the seed is weathered or known to carry seed-borne diseases that are amenable to control by ordinary seed treatment practices.

Sowing the Crop

Rye should be sown in time to make considerable growth before winter sets in. Good crops are obtained if seed is sown in late August in Canada or the northern part of the United States. The date may be later depending upon the latitude and whether or not the crop is to be used for fall pasture. Nuttonson (1958) showed that rye was planted in early August at a north Finland location. Harvest was in September or 13 months later.

In South Dakota Hume *et al.* (1926) in an 8-yr test found September 15 to be the optimum planting date. Champlin (1927) in

Saskatchewan after a 3-yr test concluded that the optimum planting date was September 1.   Rye may be sown somewhat later than winter wheat and will produce enough growth to live over winter.

It is preferable to sow rye with a drill at a rate of about six pecks per acre.   Depth of seeding may vary from 1 to 3 in.   Champlin (1927) seeded rye with a single disk drill at a depth of 2 to 3 in. in grain stubble.   He used the Dakold variety in his tests.   In a seeding rate test where he planted 2 to 8 pecks per acre at 1-peck intervals, he obtained the following net yields in bushels per acre respectively: 28.5, 32.2, 32.7, 35.2, 35.2, 35.4, and 36.2.   The highest net yield was obtained at the 8 pecks per acre seeding rate but this probably was not significantly different from the 5-peck rate.   When sowing between cotton rows in southern United States, the rate may be as low as 2 to 4 pecks per acre.

Harvesting

In areas where rye is widely grown, it matures earlier than other small grain crops.   A large part of the crop is harvested by the use of a combine.   However, there is still a large portion that is cut by a binder and later threshed by a stationary thresher.   Rye tends to shatter and should be combined as soon as moisture is low enough for safe storage.   It may be higher in moisture content if cut with a binder.   In this case shocks should be made so as to protect grain from rains and dews.

Storing

Rye is stored in many types of buildings and under a wide variety of conditions.   On the farm, most buildings are wood in construction and frequently are divided for livestock use and grain storage. Round metal bins and Quonset huts may be used for farm storage in the north central region of the United States.   Concrete bins and metal bins are used commercially.   Shedd and Cotton (1949), as well as Phillips and Hansen (1954), noted that all types of bins should meet the following basic requirements: (1) retain quality of grain; (2) exclude forms of water; (3) protect against thieves, rodents, birds, poultry, insects, and objectionable odors; (4) provide for effective fumigation; (5) be safe from fire and wind and; (6) have adequate head room for sampling.

Technical studies of grain concerning respiration, gas interchange, and insect and moisture relations have been undertaken with wheat, but infrequently with rye.   Since the structure of the rye kernel is similar to that of wheat in many respects, it is assumed that storage conditions satisfactory for wheat would prove equally satisfactory for rye. Peters *et al.* (1964) in storage experiments showed that rye could be stored 8 yr at 6% humidity in vacuum and still retain high germination capacity with relatively good vigor.

Since rye matures earlier in the summer, the moisture content more quickly reaches a safe storage level when compared to wheat or other grains. The first means of avoiding spoilage is to store rye at a satisfactory moisture level. This varies somewhat with geograph ical location. If harvested when moisture is near 13%and stored-under dry conditions free from insects, rye should remain free from storage trouble. Stored rye should be examined after binning in order to follow temperature changes. If heating is noted, the grain should be recleaned and moved to another bin. Forced ventilation can be used commercially.

Nuttonson (1958) presented a chart showing the moisture content of rye at harvest time in various parts of the Soviet Union. A high proportion of the rye is grown in areas where the moisture at harvest is above 15%. If stored at such moistures in the United States, undoubtedly there would be considerable trouble. The outside temperatures in Russia very probably are low enough to prevent spoilage in the early fall and winter. A large portion of the rye grown in Europe is harvested under rather damp conditions which results in sprouting of the seed, as suggested |by Hintzer and de Miranda (1954).

Walkden *et al.* (1954) listed the following precautionary measures to avoid grain storage troubles: (1) Store grain in a well constructed, isolated granary. (2) Store the grain in as dry a condition as possible. (3) Remove all old grain from bins and any grain and feed accumulations from other buildings on the farmstead to prevent a buildup of insect populations. (4) Apply residual spray to the ceilings, walls, and floors of the granary or crib and other buildings at least two weeks before feed grain is to be stored. (5) Fumigate all old grain which cannot be removed from the granary before new grain is binned. (6) Fumigate unprotected small grains within six weeks after harvest. (7) Inspect grain at frequent intervals to discover insect infestations or heating. (8) Fumigate the binned grain a second time if infestations develop.

CHEMICAL AND PHYSICAL CHARACTERISTICS

Chemical Analysis

Only a small amount of work has been done making strict comparisons between rye varieties except for plot yields. Instead, com-

TABLE 34

CHEMICAL COMPOSITION OF RYE GRAIN, FLOUR, AND MIDDLINGS COMPARED WITH WHEAT AND BARLEY[1]

Product	Total Digestible Nutrients	Protein	Fat	Fiber	Nitrogen-free Extract	Minerals			
						Total	Ca	P	K
					%				
Rye grain	76.5	12.6	1.7	2.4	70.9	1.9	0.10	0.33	0.47
Rye flour	74.5	11.2	1.3	0.6	74.6	0.9	0.02	0.28	0.46
Rye middlings	72.0	16.6	3.4	5.2	61.2	3.8	0.06	0.63	0.63
Wheat:									
Avg. all types grain	80.0	13.2	1.9	2.6	69.9	1.9	0.04	0.39	0.42
Barley grain, common—not including Pacific Coast States	77.7	12.7	1.9	5.4	66.6	2.8	0.06	0.40	0.49

[1] From Morrison (1956).

posite samples of rye have been examined for chemical and physical characteristics. Morrison (1956) gave the overall chemical compositions of rye grain, flour, and middlings as shown in Table 34. He also compared the chemical composition of wheat and barley. Protein and fat composition of rye, wheat, and barley grains were generally similar, ranging close to 13 and 1.8%, respectively. Fiber percentages of wheat and rye grain are essentially similar as is true of nitrogen-free extract and minerals. Rye was somewhat lower in total digestible nutrients than barley or wheat. Total digestible nutrients of rye flour and middlings were lower than that of the grain. The flour was lower than rye grain in protein, fat, fiber, and minerals. However, middlings had higher protein, fat, fiber, and mineral percentages, but lower nitrogen-free extract than either the grain or flour.

Schuette and Palmer (1938) analyzed rye germs and found that they contained 13.23% ether extract; 39.76% crude protein; 27.37% carbohydrates; 6.82% lignin; and 2.44% crude fiber. Albuminoid nitrogen was the largest component of the protein. Sucrose, pentosans, starch, and raffinose in decreasing percentages made up the carbohydrate content. These workers stated that rye germ is a high protein, phosphorus-rich substance in which the lipoids predominate over the phytin forms of the latter.

Koz'nima et al. (1956) demonstrated the presence of wedge and adhesive proteins in the rye endosperm according to Voneš et al. (1964 A). The latter workers observed 8 or 9 migrating components in a moving boundary electrophoresis study of wedge proteins in rye flour. Gluten from these wedge proteins gave essentially the same mobility values. With preparation of rye flour lacking wedge proteins these workers (1964B) produced bread with dry, crumbling, and cracking crumbs that lacked coherence. This demonstrated the need of wedge proteins for formation of normal dough. They believed that wedge proteins and gums did not interact for increased viscosity. Higher than expected viscosity values were observed in solutions of gums and glutens, thus suggesting complex formations. Korkman and Linko (1966) placed much emphasis on the role of starch in making good loaves and gave less credit to the interaction between soluble high molecular gums with protein factions, including wedge proteins. Schopmeyer (1962) stated that fine particles separated by air had 14.4% protein, while coarse particles contained only 7.3% protein.

Rohrlich and Rasmus (1956) compared the proteins of rye and wheat after manually separating small quantities of the germ, aleurone, and endosperm.   In the rye aleurone (and pericarp) they found almost equal quantities of albumin, globulin, and prolamine. In contrast, wheat had a much higher proportion of albumin in the aleurone and a much lower proportion of prolamine.   The rye was a tetraploid form, presumably Tetra Pekus.   Paper chromatography was used in determining the amino acids present in the aleurones, germs, and meal or flour from wheat and rye.   They were able to identify 18 amino acids as follows:

Alanine	Histidine	Proline
Arginine	Leucine	Serine
Aspartic Acid	Lysine	Threonine
Cystine	Isoleucine	Tryptophan
Glutamic Acid	Methionine	Tryosine
Glycocoll	Phenylalanine	Valine

These workers did not find qualitative differences between the amino acids of germ and the aleurone.   They noted, however, that the wheat endosperm was lower in arginine and lysine when compared to that of the rye endosperm.

Orr and Watt (1957) summarized the data available on 14 amino acids expressed in terms of amino acid content per gram of nitrogen present in whole grain and flours of different extractions as well as amino acid content for 100 gm of bread from light and medium rye flours.   More amino acid content was found in bread from medium flour than bread made from light flour.   Of the amino acids listed above, they did not present data for alanine, aspartic acid, glycocoll, and proline.   Glutamic acid, leucine, valine, arginine, and phenylalanine were among those that had the highest percentages.

In studying higher molecular gums of cereals, Preece and Hobkirk (1953) found that rye was rich in water-soluble pentosans with very little if any contamination with beta-glucosan.   Pure pentosans were more readily obtained from rye than from other cereals.   The major units of gums identified from rye were glucosan, xylan, araban, and galactan.   These workers (1955) later separated rye and oat polysaccharides by electrophoresis.   Rye had clear separation into two bands, while oats had three bands.   They were able to recognize the following sugar units in hydrolysates of rye: glucose, xylose, galactose, and arabinose.

Physical Properties

The physical properties of the rye grain are concerned with bushel weight, kernel weight, and other characteristics.   Though the legal weight for rye is 56 lb per bushel, only plump rye of the diploid type frequently reaches this weight.   Tetraploid rye is about two pounds less in bushel weight.   The kernel weight of rye is about 26 mg while the tetraploid forms are nearer 40 mg per kernel.   As already suggested, rye kernels vary a good deal in color and most of them tend to be "rough" in appearance because of wrinkled pericarp (Fig. 25). Further, rye tends to sprout before harvest according to many European reports.   Rye samples with ruptured pericarp above the embryo are frequently observed; this suggests incipient sprouting.

Ljung (1948) noted differences in germination of rye varieties immediately after harvest, one reaching 22.6 % while another had only 9.0% after being placed under favorable germination conditions for 4 to 5 days.

IMPORTANT FUNCTIONAL CHARACTERISTICS

Flavor

On a world-wide basis the rye grain is used to a great extent in making bread, and to a lesser extent for feeds and distilled products. Of importance to the user is the flavor which means different things to different people.   This is especially true for the taste of rye bread.

Rye "Glutens"

Johnson and Bailey (1925) studied the gluten and gas-retaining capacity of rye flour dough.   They found that the rate of gas production in the dough was high, but that the gas-retaining capacity was low and was probably responsible for the dense compact loaves that are ordinarily baked from pure rye flour.   Cunningham *et al.* (1955) extracted "glutens" from rye, wheat, barley, and oats by use of formic, oxalic, and citric acids.   The formic acid gluten of rye absorbed 70.3% water while that of wheat, barley, and oats absorbed 65.0, 55.2, and 50.2%, respectively.   These workers thought that carbohydrate gums may have caused differences in water absorption rather than variations in the protein moiety.   A judging panel thought that rye gluten was intermediate in elasticity and cohesiveness when compared to wheat gluten which was quite elastic and cohesive.

Palatability as Feed

In feeding rye, palatability is thought to be important.   Many workers consider rye unpalatable, and it is probably this reason that prompted Delwiche *et al.* (1940) to suggest mixtures of rye in the grain portion of cattle rations. Wilson and Wright (1932) in South Dakota noted that cattle did not eat large quantities of rye when fed free choice. They further noted that the finish of the cattle lacked luster. In their feeding experiments with hogs, daily gains with rye were greatly improved when it was mixed with corn or barley.

QUALITY TESTS

Flour

One of the real problems facing the rye bread maker in areas where humidity is high at harvest time, such as northwest Europe, is the sprouting or incipient sprouting of grain prior to harvest. Teden (1964) noted that a low amylase selection could tolerate one week longer in bad harvest weather without loss of germination or other characteristics. Korkman and Linko (1966) indicated that certain enzymatic action may take place even though plumule growth is concealed. Such growth may be enough for activation of alpha amylase that breaks down starch during baking. High amylase flours tend to give sticky crumbs.

Hagberg (1961) developed a viscometer-stirrer method for determining alpha amylase activity using a flour paste. After 1 min the stirrer is allowed to drop 70 mm from the uppermost position. The amount of time in seconds is called "the falling number." High amylase material requires less time than low amylase preparations. The "falling number" or sinking time is closely related to activity expressed in SKB (Sandstedt, Kneen, and Blish) units. Patterson and Crandall (1967) showed curvilinear relationships between the more time-consuming amylograph method and the faster "falling number" method for flour evaluation with especial reference to amylase activity.

Bread Making

Strict quality tests for variety comparisons are relatively unavailable for rye. Hintzer and de Miranda (1954) studied the baking quality of 8 varieties of diploid rye grown at 6 locations in The Netherlands in 1948. They tabulated their results for the field means of six varieties, and noted that the location where grown had a great influence on the baking quality. The differences among fields seemed to be greater than among varieties. Apparently the grain from the field identified as "F" had sprouted and therefore had greater alpha amylase activity and soluble substance in the bread. Field "E" had greater viscosity as measured by Brabender units and

TABLE 35

CHEMICAL, MILLING, AND BAKING RESULTS ON TETRA PETKUS AND OTHER VARIETIES OF RYE GROWN IN WISCONSIN AND PENNSYLVANIA IN 1954 AND 1955[1]

Kind of Test and Unit	Madison, Wisc. 1954			Composite from Madison, Racine, Ashland and Marshfield, Wisc. 1955		Composite from Penna. State University & Landisville, Pa. 1955		
	Adams	Caribou	Tetra Petkus	Tetra Petkus	Adams	Tetra Petkus	Balbo	von Rumker
Protein,[2] %								
Grain	11.2	11.4	14.4	10.9	9.8	11.0	12.8	10.8
Flour	6.8	9.3	11.3	7.7	6.6	8.0	9.6	7.4
Weight per kernel, mg				41.8	26.1	43.7	27.2	34.7
Flour ash,[2] %	0.67	0.44	0.48	0.58	0.58	0.63	0.64	0.65
Diastatic activity,[3] mg	224	220	261	201	217	180	165	162
Test wt per bu,[4] lb	56.0	56.9	52.1	52.5	52.1	54.5	56.2	55.5
Flour yield,[5] %	62.8	61.9	61.9	57.9	60.1	62.2	57.0	60.8
Amylograph (flour) values	...	...	...	167	292	340	850	565
Mixogram pattern	...	...	...	Md. weak	Weak	Weak	V. Weak	Weak
Bread[6]								
Blend of 40% rye and 60% first clear wheat flour								
Absorption, %	59.0	60.0	61.0	60.0	58.5	59.0	62.0	63.0
Loaf volume, ml	662	687	678	659	645	651	656	694
Grain and texture, score	70 G	65 G	50 G	90	85	70 G	90 VG	95 VG

[1] Fifield and Reitz (1958).
[2] 14.0% moisture basis.
[3] Mg of maltose per 10 gm of flour.
[4] Dockage free.
[5] Moisture free basis.
[6] Formula ingredients—flour, salt, yeast, shortening, and water.
Symbols Used: VG, very good; G, good.

panimeter value which is a measure of the bread in recovering its original volume after being compressed. The variety listed as "I" had greater amylase activity which was reflected by reduced viscosity. Bread from rye of variety "II" had more compression resistance. In addition, these workers compared diploid and tetraploid rye flours for baking qualities. They photographed 3 loaves of tetraploid rye bread that appeared to be satisfactory for quality, and 1 loaf of diploid rye bread that showed a large air hole and excessive starch degradation. The latter was attributed to more sprouting in the diploid rye. They further noted low amylase activity of tetraploid rye which is in contrast to the report of Fifield and Reitz (1958) who found that the average amylase activity of Tetra Petkus was higher than that of the comparably grown diploid varieties Adams, Balbo, and von Rumker. The latter workers made baking tests and physical and chemical tests of the rye grain, and its flour. They provided test weights per bushel, kernel weights, diastatic activity, amylograph values for the flour as well as other comparisons as shown in Table 35. They blended rye flour 40% with 60% first clear wheat flour and made baking tests with this mixture. They found that Tetra Petkus flour made bread that was generally satisfactory and of about the same quality as the bread of diploid rye samples. Bread from Tetra Petkus rye scored lower in grain and texture than bread from the diploid ryes and further, Tetra Petkus dough samples were sticky and handled with difficulty. Taste panels, as might be expected, did not agree as to flavor preference. In 1954, 12 of 15 tasters indicated that the flavor of Tetra Petkus bread was stronger than that of the 3 diploid ryes. In 1955, 9 persons of a 20-member team found the bread from Tetra Petkus stronger and more pleasing, while 5 preferred the flavor of the bread from the Adams variety, and 6 could find no flavor difference between the bread of the 2 types of rye.

Vitamins

Ihde and Schuette (1941) analyzed commercially-made rye flour for its content of the following B vitamins: thiamine, nicotinic acid, riboflavin, and pantothenic acid. They reported detailed figures for whole rye, rye germ, and other milled products. Their results are given in Table 36. Three of the vitamins had their greatest concentration in the germ, while pantothenic acid reflected higher values for the outer layers of the kernel. Booher *et al.* (1942) summarized findings on vitamin A and B_1 values. They reported the lower percentage extraction flour from Germany to be lowest in vitamin con-

TABLE 36

VITAMIN B VALUES FOR RYE AND SOME OF ITS PRODUCTS[1]

Product	Thiamine	Individual B Vitamin Readings Nicotinic Acid	Riboflavin	Pantothenic Acid
	(μg/Gm)	(μg/Gm)	(μg/Gm)	(μg/Gm)
Whole rye	2.4	12.9	1.5	10.4
Rye germ	9.3	27.0	4.46	13.9
Middlings from whole rye degermed	3.3	16.7	2.5	23.1
Middlings from degermed rye	3.3	17.7	2.0	23.1
Dark flour from whole rye	3.2	12.2	1.7	13.4
Dark flour from degermed rye	3.6	12.5	1.8	14.9
White rye flour from whole flour (not bleached)	1.5	7.1	0.76	7.1
White flour from whole rye (bleached)	1.6	7.3	0.69	7.25
White flour from degermed rye (bleached)	1.4	7.3	0.68	6.5

[1] After Ihde and Schuette (1941).

tent while it was higher as the greater per cent of the grain was milled. Schultz *et al.* (1941) found that rye varieties had different levels of B_1 vitamin but that these differences probably were not significant.

Nutrition

Sure (1954) fed the Wistar strain of albino rats whole wheat and whole rye flours that were adjusted to protein levels of 9, 8, and 5%. The rye was grown in Arkansas and the wheat was supplied by General Mills. At all levels of protein the animals made considerably more growth and showed greater protein efficiency on whole rye flour than on whole wheat flour. Increased growth in favor of rye was 39.4, 63.5, and 177.4% for 9, 8, and 5% protein levels, although gains were reduced at the lowest protein percentage. Later Sure *et al.* (1957) essentially confirmed these studies.

Whole rye flour gave more gains than whole wheat flour, and gains were much greater than for millet, corn, or sorghum. When 5% fish flour was added to the ration, rye was no longer superior to wheat or sorghum. Kihlberg and Ericson (1964) used four rye flours of 98% extraction in rat feeding studies and observed higher gain in weight than wheat of 70% extraction. They noted that rye flour had a ratio of essential to nonessential amino acid higher than wheat flour. In their test lysine was the most limiting amino acid for growth, yet low supplementation levels gave maximum response. Threonine was the next most limiting. Mixtures of several other amino acids gave weight gains. Nonsupplemented rye flour gave better gains then lysine-supplemented wheat flour. Rye crisp had a low protein value, evidently having lost nutritive value in baking. All of the rye flours were similar in amino acid content except for lysine which varied from 2.9 to 4.1 gm/16 gm nitrogen. The authors made no comments on the fact that one flour had 2.03% nitrogen and another as low as 1.47. Presumably the higher nitrogen flour would also have higher amino acid content. Ihde and Schuette (1941) found that flour contained $2/3$ as much of the total vitamin B as the whole grain and that it was definitely a richer source of the B complex than whole wheat. Even refining the rye flour failed to reduce the vitamins to as low a level as that of wheat flour.

BIBLIOGRAPHY

ALBERT, A. R. 1951. Better crops and incomes from sandy soils. Wisconsin Univ. Agr. Extens. Ser. Circ. *402*.

ANTROPOV, V. I., and ANTROPOV, V. F. 1948. Synopsis of the species and varieties of the genus Secale. Rev. Arg. Agron. *15*, 33–52.

BANNEICK, A. 1962. Comparisons of performance and experiments on cultural technique with tetraploid and diploid rye. Albrecht-Thaer-Arch. *6*, 307–321.

BOOHER, L. E., HARTZLER, E. R., and HEWSTON, E. M. 1942. A compilation of the vitamin values of foods in relation to processing and other variants. US Dept. Agr. Circ. *638*.

BROWN, H. M., and NELSON, L. V. 1957. Tetra Petkus rye. Mich. State Univ., Dept. Farm Crops. Mimeo. Circ. *22.1*.

CALDWELL, R. M. *et al.* 1958. Small grain varieties for Indiana. Recommendations for 1958 and performance, 1953–57. Purdue Univ. Agr. Expt. Sta. Research Bull. *658*.

CHAMPLIN, M. 1927. Rye production in Saskatchewan. Saskatchewan Univ. Agr. Extens. Bull. *35*.

COLLIER, G. A. 1949. Grain production and marketing. US Dept. Agr. Misc. Publ. *692*.

CUNNINGHAM, D. K., GEDDES, W. F., and ANDERSON, J. A. 1955. Preparation and chemical characteristics of the cohesive proteins of wheat, barley, rye and oats. Cereal Chem. *32*, 91–106.

DELWICHE, E. J., ALBERT, A. R., and BOHSTEDT, G. 1940. Winter rye, growing and feeding. Wisconsin Univ. Extens. Ser. Circ. *301*.

DEODIKAR, G. B. 1963. Rye. *Secale cereale* Linn. Indian Council of Agricultural Research, New Delhi. Cereal Crop Series No. III, 1–152.

DRISCOLL, C. J., and ANDERSON, L. M. 1967. Cytogenic studies of Transec—a wheat-rye translocation line. Can. J. Gen. and Cytol. *9*, 375–380.

DUMON, A. G., and LAERMANS, R. 1963. The inheritance of chlorophyll in *Secale cereale* L. Agricultura (Louvain) *11*, 91–105.

FERWERDA, F. P. 1956. Recurrent selection as a breeding procedure for rye and other cross-fertilized plants. Euphytica *5*, 175–184.

FERWERDA, F. P. 1962. Results of one cycle of recurrent selection in rye Euphytica *11*, 221–228.

FIFIELD, C. C., and REITZ, L. P. 1958. A report on Tetra Petkus, a tetraploid strain of rye. US Dept. Agr. ARS *34–7*.

FLAKSBERGER, C. A. *et al.* 1939. Key to true cereals, wheat, rye, barley, oats. The People's Commissariat of Agriculture of the USSR. Lenin Mem. All-Union Acad. Agr. Sci., Inst. Plant Cult.

FRÖST, S. 1963. Number of accessory chromosomes and protein content in rye seeds. Hereditas *50*, 150–160.

FRÖST, S. 1966. Significant differences in phenolic compounds of inbred lines. Hereditas *55*, 68–72.

FRÖST, S., and ELLERSTRÖM, S. 1965. Protein content and fertility in tetraploid rye. Hereditas *54*, 119–122.

GRAFIUS, J. E. 1951. Pierre rye. S. Dakota Agr. Expt. Sta. Bull. *406*.

HAGBERG, S. 1961. Note on a simplified rapid method for determining alpha amylase activity. Cereal Chem. *38*, 202–203.

HANCOCK, N. I., and OVERTON, J. R. 1960. Behavior and adaptation of Balbo and Tetra Petkus rye. Tenn. Agr. Exp. Sta. Bull. *307*.

HARRINGTON, J. B. 1953. Three new varieties: Antelope winter rye, Husky barley and Torch oats. Saskatchewan Univ. Field Husbandry Dept. Circ. *557*.

HERIBET-NILSSON, N. 1916. Population analyses and experiments on self-sterility, self-fertility, and sterility in rye. Z. Pflanzenzücht. *4*, 1–41.

HILPERT, G. 1957. Effect of selection for meiotic behaviour in auto-tetraploid rye. Hereditas *43*, 318–322.

HINTZER, H. M. R., and DE MIRANDA, H. 1954. Investigations on the quality of diploid and tetraploid rye for breadmaking. Cereal Chem. *31*, 407–416.

HUME, A. N., HARDIES, E. W., and FRANZKE, C. 1926. The date of seeding winter rye when the ground is dry or wet. S. Dakota Agr. Expt. Sta. Bull. *220*.

IHDE, A. J., and SCHUETTE, H. A. 1941. Thiamine, nicotinic acid, riboflavin and pantothenic acid in rye and its milled products. J. Nutr. *22*, 527–533.

JASNY, N. 1940. Competition Among Grains. Food Res. Inst., Stanford Univ., Palo Alto, Calif.

JOHNSON, A. H., and BAILEY, C. H. 1925. Gluten of flour and gas retention of wheat flour doughs. Cereal Chem. *2*, 95–106.

KIHLBERG, R., and ERICSON, L.-E. 1964. Amino acid composition of rye flour and the influence of amino acid supplementation of rye flour on growth, nitrogen efficiency ratio and liver fat in the growing rat. J. Nutr. *82*, 385–394.

KHUSH, G. S. 1963. Cytogenetics of weedy ryes and origin of cultivated rye. Econ. Botany *17*, 60–71.

KOO, F. K. S. 1958. Deleterious effects from interpollination of diploid and autotetraploid winter rye varieties. Agron. J. *50*, 171–172.

KORKMAN, M., and LINKO, P. 1966. Activity of different enzymes in relation to the baking quality of rye. Cereal Sci. Today *11*, 444–447.

KOZ'NIMA, N. P., IL'INA, V. N., and BUTMAN, L. A. 1956. Gluten proteins of rye grain. Doklady Akad. Nauk. SSSR *110*, 610–612.

KRANZ, A. R. 1957. Genetic analysis of primitive ryes from Iran. Z. Pflanzenzücht. *38*, 101–146.

KRANZ, A. R. 1963. Contribution to cytological and genetical research on evolution in rye. Z. Pflanzenzücht. 50, 44–58.

KUCKUCK, H., and KRANZ, A. R. 1957. A genetic analysis of rye populations from Iran. Wheat Information Ser. Circ. No. 6, 20–21. Kyoto Univ., Kyoto, Japan.

LEIGHTY, C. E. 1916. Culture of rye in the eastern half of the United States. US Dept. Agr. Farmers' Bull. 756.

LEITH, B. D., and SHANDS, H. L. 1938. Fertility as a factor in rye improvement. J. Am. Soc. Agron. 30, 406–418.

LJUNG, E. W. 1948. The rye breeding work of the seed association. In AKERMAN, A., TEDIN, O., and FROIER, K. Svalof 1886–1946, History and Present Problems. Carl Bloms Boktryckeri A.-B. Lund, Sweden.

LUNDQUIST, S. 1966. Heterosis and inbreeding depression in autotetraploid rye. Hereditas 56, 317–366.

MARTIN, J. H., and SMITH, R. W. 1923. Growing rye in the western half of the United States. US Dept. Agr. Farmers' Bull. 1358.

MCBEAN, D. S. 1966. Frontier—a new variety of winter rye. Can. J. Plant Sci. 45, 398–399.

MOOERS, C. A. 1933. Balbo rye. Tennessee Univ. Agr. Expt. Sta. Circ. 45.

MOREY, D. D. 1964. Wesser rye—a new variety for the Southeast. Georgia Agr. Res. 6, 10–11.

MORRISON, F. B. 1956. Feeds and Feeding, 22nd Edition. The Morrison Publishing Co., Ithaca, N.Y.

MÜNTZING, A. 1939. Studies on the properties and the ways of production of rye-wheat amphiploids. Hereditas 25, 387–430.

MÜNTZING, A. 1954. An analysis of hybrid vigor in tetraploid rye. Hereditas 40, 265–277.

MÜNTZING, A. 1962. Effects of accessory chromosomes in diploid and tetraploid rye. Hereditas 49, 371–426.

MÜNTZING, A. 1963. A case of preserved heterozygosity in rye in spite of long-continued inbreeding. Hereditas 50, 377–413.

NAKAJIMA, G. 1962. Cytological studies on interspecific hybrids of rye. IV. External characteristics and PMC meiosis in F_1 hybrids between S. ancestrale and 6 other species of rye. Senshokutai (Chromosome) Kromosomo 1962, No. 51–52, 1675–1683.

NUTTONSON, M. Y. 1958. Rye-climate Relationships and the Use of Phenology in Ascertaining the Thermal and Photo-thermal Requirements of Rye. American Institute of Crop Ecology, Washington, D. C.

ORR, M. L., and WATT, B. K. 1957. Amino acid content of foods. US Dept. Agr. Home Econ. Res. Rept. No. 4.

PATTERSON, B. E., and CRANDALL, L. G. 1967. Amylograph vs. falling number values compared. Cereal Sci. Today 12, 332–335.

PATTERSON, F. L., and MULVEY, R. R. 1954. Tetra Petkus rye. Purdue Univ. Agr. Expt. Sta. Mimeo. Circ. AY-71A.

PETERS, C., SCHNEIDER, E., and TOLPEL, E. 1964. Observations on the preservation of germination capability and sprouting power of breeding stocks of winter rye by storage in vacuum. Züchter 34, 135–138.

PHILLIPS, R., and HANSEN, R. W. 1954. More grain storage on Iowa farms. Iowa Farm Sci. 9, No. 1, 3–6.

PIVNENKO, M. J. 1963. Efficacy of heterosis in winter rye. News Agr. Sci. 6, 41–45.

POPOV, G. I. 1962. The method of vegetative values for breeding winter rye. Vestn. Sel'skokozgaptu. Nauk. (Rep. Agr. Sci.) 1962, No. 12, 101–106.

PREECE, I. A., and HOBKIRK, R. 1953. Non-starchy polysaccharides of cereal grains. III. Higher molecular gums of common cereals. J. Inst. Brewing 59, 385–392.

PREECE, I. A., and HOBKIRK, R. 1955. Paper electrophoresis of polysaccharides. Chem. Ind. (London) 1955, No. 10, 257–258.

PUHR, L. F. 1962. Twenty years of soil management on Vienna silt loam. S. Dak. State College Bull. 508.

ROBINSON, R. G., and KOO, K. S. 1954. Comin' through with Caribou. Minn. Farm and Home Sci. 11, No. 2, 17.

ROHRLICH, M., and RASMUS, R. 1956. Experiments on the chemical differentiation of the proteins of wheat and rye. Z. Lebensm. Untersuch. u. Forsch. 103, 89–96.

ROSHEVITZ, R. I. 1947. Monograph of the genus Secale L. Flora et systematica plantae vasculares. Series 1. Fasc. 6, 105–163. (Russian.)

ROSS, C. 1953. New lodge-defying tetraploid. Southern Seedsman 16, 16–17, 72.

RUEBENBAUER, T., and BISKUPSKI, A. 1963. Translation [Preliminary estimation of the milling and baking value of tetraploid rye.] Hodowra Roślin Aklimatyz. Nasiennictwo 7, 1–15.

SADANAGA, K. 1956. Cytological studies of hybrids involving Triticum durum and Secale cereale. Wheat Inf. Ser. No. 3, 23–24. Kyoto Univ., Kyoto, Japan.

SANCHEZ-MONGE, E. 1956. Fertility in Triticale. Wheat Inf. Ser. No. 3, 29. Kyoto Univ., Kyoto, Japan.

SCHOPMEYER, H. H. 1962. Rye and rye milling. Cereal Sci. Today 7, 138–143.

SCHUETTE, H. A., and PALMER, R. C. 1938. The chemistry of the rye germ. IV. Its proximate composition. Cereal Chem. 15, 445–450.

SCHULTZ, A. S., ATKIN, L., and FREY, C. N. 1941. A preliminary survey of the vitamin B content of American cereals. Cereal Chem. 18, 106–113.

SHANDS, H. L., and FORSBERG, R. A. 1964. Distinguishing plant characters in certain inbred rye lines. Agron. Abst. 80.

SHEDD, C. K., and COTTON, R. T. 1949. Storage of small grains and shelled corn on the farm. US Dept. Agr. Farmers' Bull. 2009.

SILVÁN, A. 1967. Breeding by induction of polyploidy in sugar beet and rye. Project No. E 25-CR-5, PL 480, FG-Sp-120. Mimeo. 68.

SNEEP, I. J., OLTHOFF, B. H., and HOGEN ESCH, J. A. 1967. Forty second descriptive list of field crop varieties. Post Box 32. Wageningen.

SPRAGG, F. A., and NICOLSON, J. W. 1917. Rosen rye. Mich. State Univ. Agr. Ext. Bull. 9.

SPRAGUE, H. B. 1938. Breeding rye by continuous selection. J. Am. Soc. Agron. 30, 287–293.

SURE, B. 1954. Protein supplementation. Relative nutritive values of proteins in whole wheat and whole rye and effect of amino acid supplement. J. Agr. Food Chem. 2, 1108–1110.

SURE, B., EASTERLANG, L., DOWELL, J., and CRUDUP, M. 1957. The addition of small amounts of defatted fish flour to whole yellow corn, whole wheat, whole and milled rye, grain sorghum and millet. I. Influence on growth and protein efficiency. II. Nutritive value of the minerals in fish flour. J. Nutr. 63, 409–416.

SYBENGA, J., and PRAKKEN, R. 1963. Gene analysis of rye. Genetica (33), 95–105.

TEDEN, O. 1964. Problems in rye breeding: The importance of methodological research. Sverig. Utsädesfören Tidskr. 74, 363–377.

TOROP, H. A., and PAHMOVA, V. P. 1966. Hybridization as a method of increasing the breeding value of tetraploid rye. Genetics 4, 114–118.

UNRAU, A. M., and JENKINS, B. C. 1964. Investigations on synthetic cereal species. Milling, baking, and some compositional characteristics of some "Triticale" and parental species. Cereal Chem. 41, 365–375.

VAVILOV, N. 1926. Studies on the origin of cultivated plants. Bull. Appl. Botany Plant Breeding (Leningrad). 16, 139–248. (English translation.)

VOLIN, L. 1951. A survey of Soviet Russian agriculture. US Dept. Agr. Monograph 5.

VONEŠ, F., PODRAZKÝ, J., ŠIMOVÁ, J., and VESELÝ, Z. 1964A. Electrophoretic properties of wedge protein and gluten of rye flour. Cereal Chem. 41, 9–15.

VONEŠ, F., PODRAZKÝ, J., ŠIMOVÁ, J., and VESELÝ, Z. 1964B. Significance of wedge protein of rye flour for dough properties. Cereal Chem. 41, 456–464.

WALKDEN, H. H., WILBUR, D. A., and GUNDERSON, H. 1954. Control of stored grain insects in the North Central states. Minn. Agr. Expt. Sta., North Central Regional Pub. 49.

WARREN, F. S., and HAYES, H. K. 1950. Correlation studies of yield and other characteristics of rye polycrosses. Sci. Agr. 30, 12–29.

WELLS, D. G. 1966. Registration of Explorer rye. Crop Sci. 6, 501–502.

WIDDIFIELD, R. B. 1953. Grow winter rye for better weed control. N. Dakota Agr. Ext. Ser. Circ. A-199.

WILLIAMS, W. R. 1952. Principles of Agriculture. Translated by G. V. Jacks. Chemical Publishing Co., New York.

WILSON, J. W., and WRIGHT, T. 1932. Rye as a fattening feed for cattle and swine in South Dakota. S. Dakota Agr. Expt. Sta. Bull. 271.

CHAPTER 6

N. W. Kramer[1]

S. A. Matz

Sorghum

INTRODUCTION

Origin and History

Sorghum is thought to have originated in Africa. That it was grown in ancient times is shown by a carving depicting the plant which was found in an Assyrian ruin dating from about 700 BC (Ball 1910).

The first seed of any kind of grain sorghum brought to the Western Hemisphere probably came in slave ships from Africa. One variety called "chicken corn" escaped from cultivation and now grows as a weed in some of the southern states; another, called "guinea corn," was once rather widely distributed but has now practically disappeared from cultivation. White Durra and Brown Durra were introduced into California from Egypt in 1874. White Kafir and Red Kafir were introduced from Natal, South Africa about 1876. Milo was introduced from Colombia about 1879, but there is no information as to how this variety reached Colombia. Shallu was introduced from India about 1890, Pink Kafir from South Africa about 1904, and Feterita and Hegari from the Anglo-Egyptian Sudan in 1906 and 1908, respectively (Vinall *et al.* 1936).

The varieties and hybrids grown in the United States today are derived mostly from milo-kafir crosses. In commerce, "milo" has been used as a synonym for sorghum grain, while "maize" or "milo maize," "kafir corn," and "gyp corn" are other terms sometimes used to identify the grain.

Recent Changes in the Crop

Grain sorghum types and varieties have changed greatly during the past 30 yr. The tall types that were grown through the 1930's have been completely replaced by shorter "double dwarf" or "combine" types, with the result that the earlier bottleneck of hand heading has been completely eliminated by combine harvesting. In the same period, earlier varieties with a consequently lower total water requirement were developed; these resulted in greater

---

[1] DR. N. W. KRAMER is Research Director, Paymaster Seeds Div. of Anderson Clayton and Co.

sureness of production in the older sorghum growing areas and in the extension of production into drier areas and into areas with shorter growing seasons. Sorghum hybrids, which first became available in 1955, made it possible to increase yields 20 to 40% and further increased the acreage of the crop by making it more profitable to producers. Improvement in hybrid quality and yield is continuing as a result of the work of experiment stations throughout the Southwest.

The acreage of sorghum has also been increased by some other factors. The crop so far has been relatively free of serious disease and insect pests. The earlier varieties have proved to be particularly well suited as a late-planted crop where wheat or other crops have been abandoned or lost to weather hazards. The occurrence of drouth periods that have reduced yields of corn more than those of sorghum have resulted in the replacement of considerable corn acreage by sorghum in the western part of the Corn Belt, particularly in the areas where the annual precipitation is less than 30 in.

Sorghums with yellow endosperm similar to that of corn are being developed; this improvement will give sorghum an appreciable vitamin A and xanthophyll content, and will put it in a better competitive position as a feed grain. A much stiffer stalk is being introduced to give better resistance to lodging, and work is also underway on improvements in seed size, threshability, and grain composition and quality (Kramer 1958).

Production and Trade Statistics

Sorghum is the chief grain food in much of Africa and parts of India, Pakistan, and mainland China. It is the world's third most important food grain, being exceeded in utilization for food only by wheat and rice. Statistics on the world production of sorghum are incomplete, because much of it is grown in regions with no reporting history, but the total area of sorghum grown for grain is thought to be more than 80 million acres (Martin and Leonard 1949). In 1938, world production of grain sorghum was estimated to be over 24,000,000 short tons of which 18,000,000 tons were consumed as human foods (Anderson and Martin 1949).

Sorghum is grown on all the continents below latitudes of 45° and on many of the islands of the East and West Indies. It is an important cereal crop in several African countries and in parts of India, Pakistan, China, Manchuria, and the United States, particularly where limited rainfall prevents successful growth of corn or rice. It is also grown to some extent in nearly all the countries in the southern half of Europe and Asia, and in Central America, South America, and Australia.

In the United States, sorghum was harvested for grain on an average of about 13,760,400 acres per year during the 10-yr period, 1956 through 1965. This is about 50% more acres than were utilized for the crop in the preceding 10 yr. Production averaged 526,377,000 bu per year during the 1956–1965 period, for an overall average yield of 38.2 bu (2,139 lb) per acre. Total production in 1967 exceeded 700,000,000 bu, indicating a continuing increase in output. As a grain crop in the United States, sorghum is currently exceeded in production only by wheat and corn, and in total acreage by corn, wheat, and oats. It is very likely to pass oats in total acreage harvested within the next decade.

The increased popularity of sorghum can be attributed to such factors as: (1) government controls on wheat and cotton which release land for sorghum growing; (2) development of hybrids with a subsequent improvement in yield of 20–30%; (3) availability of short-stalked types better adapted to mechanical harvesting; (4) increased recognition by farmers of the ability of sorghum to resist temperatures and erratic rainfall; and (5) demand for waxy varieties for use as food starch sources (though this is as yet only a small outlet for the grain).

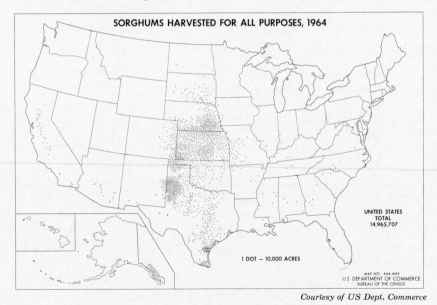

Courtesy of US Dept. Commerce

Fig. 26.   Sorghums Harvested for Grain or for Seed

Sorghum is the major feed grain grown in the Great Plains area stretching from the South Plains of Texas to South Dakota because it is better adapted to, and more productive under, the subhumid and semiarid conditions of that area than any other grain crop. Sorghum produces grain more reliably than any other crop under the hazardous farming conditions of the subhumid and semiarid areas. Its performance under adverse conditions gives it a value above the cash value of the grain and helps to maintain a more stable agricultural system. The residue from a sorghum crop that was a failure from the standpoint of grain production has often prevented the loss of land by wind erosion. Considerable acreages are also grown in the western part of the Corn Belt and in the hot irrigated areas of the Southwest, with lesser amounts in the South Central and South Atlantic States. Statistics on sorghum acreage and production are reported for 23 states (Table 37).

The leading state in sorghum production is Texas, with an average of about 43% of the US production in the 7-yr period from 1959 through 1965. The proportion of US production harvested in

TABLE 37

SORGHUM PRODUCTION[1] IN PRINCIPAL GROWING STATES

	Acreage Harvested (1,000 A)	Yield per Acre (Bu)	Production (1,000 Bu)	Sold from Farms (1,000 Bu)
Indiana	8	70.0	560	280
Illinois	10	64.0	640	294
Iowa	33	67.0	2,211	1,106
Missouri	230	57.0	13,110	5,637
South Dakota	375	30.0	11,250	4,388
Nebraska	2,329	54.5	126,930	83,774
Kansas	3,038	45.0	136,710	83,393
Virginia	9	42.0	378	42
North Carolina	43	48.0	2,064	1,032
South Carolina	6	30.0	180	34
Georgia	15	34.0	510	107
Kentucky	6	40.0	240	58
Tennessee	8	41.0	328	62
Alabama	14	26.0	364	73
Mississippi	9	35.0	315	94
Arkansas	29	35.0	1,015	609
Louisiana	5	35.0	175	52
Oklahoma	606	37.0	22,422	12,556
Texas	5,495	52.0	285,740	260,023
Colorado	368	35.5	13,064	8,492
New Mexico	225	65.0	14,625	11,700
Arizona	165	70.0	11,550	10,164
California	297	73.0	21,681	19,079
United States	13,323	50.0	666,062	503,049

[1] 1965 Preliminary, (Anon. 1966).

Texas has been declining as a result of greater increases of plantings in other states, however, sorghum is still the leading grain crop in Texas and some kind of sorghum is planted on approximately $1/3$ of the total cultivated acreage of the state. In Kansas and Nebraska, the 2nd- and 3rd-ranking states in sorghum acreage, wheat is the only cereal that exceeds sorghum in acreage. Production in Nebraska has been increasing at a very rapid rate.

In recent years there has been a great increase in sorghum plantings in other parts of the United States. This interest has been expressed in increased acreages, but the total grown outside the Sorghum Belt remains small.

The acreage of sorghum planted varies greatly from year to year. It is affected in most of the area by weather conditions at planting time and by the acreage and condition of other crops grown in the Sorghum Belt. The acreage of sorghum is inversely related to that of cotton and wheat where these crops are grown in the same area as sorghum. On the western fringe of the Corn Belt, sorghum and corn are interchangeable crops; a succession of dry years may result in considerable shifting of acreage from corn to sorghum, while a series of years favorable for corn will reduce the sorghum acreage.

The acreage of grain sorghum began an upward trend in the 1940's when combine varieties became available. The shortage of farm labor, the development of the small combine, and the wartime demand for grain all contributed to the increase. Government agricultural policies that restricted the acreages of wheat, cotton, and corn in varying amounts also affected sorghum acreages. In individual years the sorghum acreage may change considerably because of weather conditions at planting or because of factors affecting the acreage of other crops, such as winter-killing or abandonment of wheat.

Much of the large increase in sorghum production which was seen in 1956 and 1957 can be attributed to a swing to sorghum from other crops because of a succession of drouth years and because of the effect of government agricultural programs on cotton, wheat, and corn acreage; and to the availability of the new sorghum hybrids.

BOTANY

Sorghum belongs to the family Graminae, tribe Andropogonae. All the annual sorghums have ten pairs of chromosomes and belong to one species, *Sorghum vulgare*, which includes such diverse types as grain sorghums, sudangrass, broomcorn, and tall sorghums that may be grown for forage, silage, or syrup.

Sorghum is a coarse grass that may grow from about 2 to more than 15 ft in height; the height of the widely cultivated grain sorghum varieties is usually between 2 and 5 ft. The vegetative appearance of the plant is somewhat similar to corn. The stems may be fairly fine in grass sorghums or more than an inch in diameter in some grain and forage types. Some varieties of grain sorghum have juicy stalks and leaf midribs while others are dry and pithy. The leaves, of which there are usually 10 to 16, are smooth and have a waxy surface; the leaf structure is such that water loss is reduced to a low rate. The inflorescence is a loose to dense panicle bearing from 800 to 3,000 kernels. The seed of different varieties varies greatly in size, pigmentation, and other characteristics. Although all grain sorghums are annuals, they can survive as perennials where temperatures are mild.

A recent publication of considerable importance to the botanist (Murty *et al*. 1967) is based on a classification and catalog of a world collection of sorghum. It includes a description of 70 groups in the genus *Sorghum* based on a modified Snowden's classification and photographs of representative earheads of a world collection of sorghum based on a modification of the classification of this genus. An appendix contains a catalog of panicle and vegetative characters of specimens in the collection.

GROWING THE CROP

Soils and Climate

Sorghum is well adapted to semiarid areas, but it can make good use of additional water and is grown extensively under irrigation in dry areas. The most favorable mean temperature for the growth of sorghum is about 80°F. The sorghum plant withstands extreme heat better than most other crops, but extremely high temperatures during the fruiting period, in combination with low moisture levels, may reduce the grain yield.

Certain structures and processes of the plant adapt it particularly well to hot and dry conditions. The plants of some types of sorghum become partially dormant, and, as farmers say, "wait for rain" during dry, hot periods but resume growth when more water becomes available; this characteristic is particularly valuable in areas where short drouth periods may be expected to be followed by periods of rain. Types of sorghum which do not exhibit this characteristic are especially useful where rainfall is definitely seasonal and the cessation of rainfall indicates the beginning of a long dry period; under such conditions these "nonstop" types produce grain while types that tend to become dormant might produce no grain. Compared to corn, sorghum has more secondary roots, a smaller leaf area, and a markedly more xerophytic type of leaf structure (Miller 1916). Sorghum leaves inroll as they wilt and have a waxy cuticle that apparently retards transpiration.

Sorghum is grown on all types of soil; in the area in which it is most widely grown the ability of a soil to provide water is usually its most important characteristic. Under good climatic conditions sorghum does best on deep fertile soils. In dry areas sorghum often does best on sandy soils because of their water infiltration and retention characteristics. Sorghum varieties differ in their reactions to salt concentrations, soil air and moisture, and other soil properties.

Cultural practices for sorghum are generally similar to those used for other row crops such as corn and cotton. The particular procedures and implements used vary considerably, depending upon the soil, climate, preceding crop, the equipment available, and the usual practices of the particular area.

Rotation

Grain sorghum can follow almost any other crop, but in most of the sorghum growing area production hazards and limited moisture restrict the number of crops that can be grown and their arrangement into crop sequences. Grain sorghum is often grown continuously or is alternated with other crops such as forage sorghum, sudangrass, broomcorn, corn, or cotton. In some of the irrigated areas in southern California and Arizona, sorghum may be the summer crop in a double-cropping system, with wheat or barley being grown in the cool season. In a few areas with moderate rainfall, a rotation of sorghum, spring barley or oats, and winter wheat is used. In some areas where both sorghum and wheat are grown, a 3 yr rotation of sorghum-fallow-wheat is used, providing a fallow period of 9 to 11 months for the accumulation of rainfall before the planting of each crop; this rotation is particularly adapted to use in narrow strips for the control of wind erosion because some of the land is always in a growing crop and the unplanted strips are not fallowed long enough for all the residue to disintegrate. In the irrigated area of the South Plains of Texas, a rotation of sorghum, soybeans, and cotton is gaining in favor. In the Corn Belt and in most of the South, sorghum replaces corn in the cropping sequence.

Sorghum sometimes depresses the yields of succeeding crops. This is thought to be due primarily to the greater depletion of soil moisture by sorghum and to the fact that sorghum roots usually remain alive until the plant is killed by frost. Sorghum roots, a high-carbon residue, do not decay appreciably until the next spring, sometimes causing a serious deficiency of available nitrates in the early part of the following season. Early tillage and applications of nitrogen fertilizer tend to reduce this effect of sorghum.

Preparation of Land

Preparation of land for sorghum usually begins as soon as possible after the harvest of the preceding crop. In most sorghum producing areas, the saving of soil moisture and the storage of hoped-for precipitation are among the most important factors concerned in land preparation, although loosening of compacted soil and the preparation of a seedbed are also important. Soil preparation usually consists of procedures to kill weeds and other vegetation, to loosen the soil to permit more rapid infiltration of precipitation or irrigation water, to roughen or ridge the soil to catch blowing snow and reduce wind erosion, and to prepare the land for the planting operation. The common methods of land preparation are listing, subsurface tillage, plowing with either disk or moldboard plows, disking, and chiseling, or combinations of these methods. On some of the medium- to light-textured soils where the lister planter is used, the crop may be grown without any land preparation with very little reduction in yield.

Planting

Sorghum is usually planted in cultivated rows 36–44 in. apart, but considerable acreage is planted in narrower rows 20, 24, 28, or 32 in. apart or in paired rows with alternate 14- and 26-in. row spacings.

Some sorghum is planted with a grain drill but most sorghum in dry areas is planted with the lister planter or with similar equipment which pushes away the dry surface soil so that the seed can be planted in moist soil without covering it too deep. With this equipment the seed is usually planted in the bottom of a furrow. In more humid areas sorghum is planted with surface planters that leave the soil level after planting. In a few areas where drainage is a problem or where early irrigation is needed, sorghum is planted on the top of lister beds. The optimum depth of planting is about 2 in., but deeper covering of the seed is often practiced in hot dry conditions.

The amount of seed to be planted per acre depends upon many factors: the prospective grain yield; seed size; seed viability; and

Courtesy of Deere and Co.

FIG. 27.   PLANTING SORGHUM AND APPLYING FERTILIZER WITH A 4-ROW LISTER
PLANTER

conditions of soil and weather that may affect the percentage of
emergence.   Because of all the variables involved, it is impossible
to make definite recommendations of rates of seeding.   A good
rule-of-thumb for average conditions is to plant about $1\frac{1}{2}$ lb of
seed for each 1,000 lb of grain yield expected.   The application of
this rule results in recommendations of as low as 1–2 lb per acre for
some dry land conditions and as high as 10 lb per acre for favorable
conditions, all of which are within the range of seeding rates used.
Stands of sorghum can deviate considerably from optimum plant
populations with relatively little effect on grain yields (Kramer
1957).

The minimum temperature for germination of sorghum seeds
varies from 45° to 50°F.   Germination and early growth is slow at
soil temperatures below 60°F, so the planting dates for sorghum,
because of temperature requirements, are necessarily a little later
than those for corn and about the same as those for cotton.   An
equally important consideration where the growing season is long
is that of having the plant grow in the most favorable part of the
year in respect to temperatures and precipitation.   For example,
in all of Texas below the High Plains, planting is done as early as
possible so that the crop will be growing in the early part of the

summer when temperatures are lower and the probability of getting rains is greater; the latter part of the summer is hotter and drier and is much less favorable for sorghum production. In the northern part of the Sorghum Belt and at higher elevations, sorghum must be planted as early as possible because the whole season is needed for growing the crop. In the South Plains of Texas the optimum planting period is June 10 to 25, later than that of the northern areas; this date of planting results in the plants being small and having a low water use during the hottest and driest part of the summer in July and early August, and the period of highest water use by the plant is then reached in late August and early September when temperatures are lower and the probability of rainfall has increased.

Fertilizer

Sorghum uses relatively large amounts of the common fertilizer elements for high yields (Table 38). In much of the sorghum area, however, the crop is grown on dryland and yields are not high enough to be limited by fertilizer elements, so the use of fertilizer on sorghum is limited for the most part to the areas with higher rainfall and to irrigated areas.

TABLE 38

FERTILIZER ELEMENTS IN SORGHUM PLANT PARTS AT A YIELD LEVEL OF 100 BUSHELS PER ACRE[1]

Part of Plant	Production (Lb)	Nitrogen (Lb)	Phosphoric Acid (Lb)	Potash (Lb)
Grain	5,600	100	32	17
Leaves	1,770	15	5	9
Stems	4,980	10	4	14
Roots	2,920	9	10	5
Total above ground	12,350	125	41	40

[1] Quinby et al. 1958.

If fertilizer is applied, it should be available in adequate quantities when it is needed by the plant; the time of application may be very important when fertilizers are side-dressed. The uptake of nitrogen by sorghum is very high during two periods: the period of rapid vegetative growth preceding heading, and during the period of grain development. Phosphorus accumulation is highest during the period of rapid vegetative growth preceding heading (Wagner et al. 1913).

Irrigation

Although sorghum is a drouth-tolerant crop, it gives a good return from supplemental irrigation in most areas. The amount of water used by the plant is not a fixed value; it is affected by temperature, humidity, wind movement, soil moisture tension, and the duration of growth of the variety.

The most effective use of water is obtained when the crop is provided with adequate water for continuous vigorous growth throughout the season. In average conditions in the southern part of the Sorghum Belt the total consumption of water is usually 21–23 in. for a 100-bu grain crop. Whenever the crop is in severe moisture stress, that is, if the available moisture in the top 2 ft of soil is reduced below 25% of the total available storage capacity, potential yields are reduced.

Water use is low during the early growth of the sorghum plant, usually averaging 0.05–0.10 in. per day for the first 2–4 weeks. Peak use usually occurs during the late boot and early heading stages and may reach 0.33 in. per day. The water use usually averages about 0.25 in. per day from the boot through the dough stage (Swanson and Thaxton 1957). (The figures given above apply to stands of 80 to 120 thousand plants per acre with available soil moisture above 25%.)

When preplanting rainfall is insufficient to bring the soil profile to field capacity to depths of 5–6 ft, a preplanting irrigation is usually made. If the soil is wet to such depths the later irrigations are then based on the moisture content of the top 2 ft of soil. Usually the first irrigation is made 30–40 days after planting unless rainfall has occurred, and subsequent irrigations are made as necessary to maintain available water in the top 2 ft above a minimum of 25–50 %. The frequency of application will depend on the water-holding capacity of the soil and factors that affect the rate of water use.

Cultivation

Sorghum is usually cultivated as are other row crops. Cultivation is practiced primarily for weed control. If the crop is planted on land free of growing weeds, cultivation is required only as new weeds grow. Early cultivations are often made with a harrow, rotary hoe, or knife sled; later cultivations are made with conventional row cultivators. Narrow-row plantings are often cultivated entirely with a harrow or rotary hoe.

Sorghum roots grow near the surface, so cultivation must be relatively shallow because destruction of roots at any time is detrimental to the crop.

Weeds in sorghum can be controlled with any of several chemical herbicides, but caution must be exercised so that neither the sorghum crop nor other crops nearby are damaged by the chemical.

Insects and Diseases

Sorghum is attacked by a large number of insects and diseases, but they seldom cause general losses, although some may be severe in local areas.

The most destructive insect pests are the chinch bug, sorghum midge, cornleaf aphid, corn earworm, and sorghum webworm. The chinch bug is a sucking insect that often moves to sorghum from maturing small grain fields; a number of varieties are relatively resistant to this insect. The sorghum midge is a problem in the southern states; the larvae hatch within the glumes and prevent seed development. Aphids occasionally cause some damage, but other insects attacking sorghum cause only minor or local damage.

The most important diseases of sorghum can be grouped into seed and seedling diseases, leaf diseases, smuts, and root and stalk rots. Species of at least seven genera of fungi may attack seeds and seedlings; damage from this group can be greatly reduced by seed treatment with an effective fungicide and by planting when conditions are favorable for rapid seedling emergence. Leaf diseases, including at least 3 bacterial and 8 fungal diseases, are most common in humid areas; control measures are the use of resistant varieties and sanitation. Sorghum is affected by 3 smuts: 2 kernel smuts that can be controlled by seed treatment, and head smut which is soil-borne and can be controlled only by the use of resistant varieties. The most serious root and stalk rots are milo disease and charcoal rot. Periconia root rot or milo disease is controlled by planting immune varieties. Charcoal rot is often a problem when the crop is subjected to extreme heat or drought during the fruiting period; the disease causes disintegration of the pith and vascular structures in the lower part of the stalk, followed by premature drying of the plant and lodging. No complete control for charcoal rot is known, but the development of resistant varieties offers the best hope for control. Several other stalk rots are of lesser importance (Quinby *et al.* 1958).

Harvesting and Storage

Grain sorghum is usually harvested from standing stalks with a combine. The grain is physiologically mature when the greenest seeds drop to 35% moisture, but it should not be harvested until

Courtesy of Allis Chalmers Mfg. Co.

FIG. 28. HARVESTING COMBINE GRAIN SORGHUM WITH A 4-ROW SELF-PRO-
PELLED COMBINE

the grain has dried to 13% or less moisture unless the grain is to be dried artificially.

In the threshing operation the combine should be adjusted so that all the grain is threshed and separated from the stalks, a minimum amount of grain is cracked, and as little trash as possible is left in the grain.

Varieties differ in their suitability for harvesting from green plants before a killing freeze. Some varieties have a genetically conditioned head-drying character that results in the rapid drying of the grain and head, while others tend to keep the grain at moisture levels too high for harvest until the plants are killed by frost.

Chemicals can be used to kill sorghum head tissues and speed drying, thus permitting earlier harvest, but the use of these harvest-aid chemicals must be limited to fields for seed production until they are cleared for use on food and feed grains.

Most of the sorghum grain is taken directly from the combine to country and terminal elevators, but in some areas a part of the crop is stored on farms.

In the more humid areas, much of the grain must be dried before it can be stored safely. At elevators most drying is done with heated air in relatively high-capacity driers, but sorghum can be dried with unheated air quite satisfactorily. For drying with unheated air the recommended depths of grain of different moisture contents, the airflow per bushel, and the periodicity of aeration will vary with local temperature and humidity values.

A tight structure is necessary to protect stored grain from weather, birds, rodents, and insects. Bins and storage areas should be cleaned and sprayed with a residual insecticide before grain is stored. The grain should be checked for temperature, moisture, and insect activity as often as necessary under the particular storage conditions.

Stored sorghum grain may be attacked by the common insects of stored grain such as the Angoumois grain moth, rice weevil, granary weevil, and others. The storage of clean dry grain in clean bins, along with aeration and fumigation as necessary, will control these insects.

STRUCTURE AND COMPOSITION

The caryopsis or kernel of sorghum is spheroidal, and has typical dimensions of about 4.0 mm by 2.5 mm by 3.5 mm. Weight will vary from 8 to 50 mg with an average of 28 mg (Watson 1967).

Common sorghums have tan, red, or brown hulls. White hulled sorghum, commonly called kafir, generally has smaller kernel size than most other sorghums, and the regular (nonwaxy) type is seldom seen in trade. Most waxy varieties are also white hulled.

A microphotograph of a section from a steeped sorghum kernel is reproduced in Fig. 29. Some cell structure has been lost from the central endosperm but all important gross features are clearly visible. As is generally the case with cereal grains, the caryopsis can be seen to have three readily differentiated parts, the outer covering or pericarp, the large central mass or endosperm, and the germ, which is displaced to one side and end. Based on weighing of hand-dissected fractions, Hubbard et al. (1950) reported the proportion of endosperm in the varieties examined to range from 80.0 to 84.6%, the germ from 7.8 to 12.1%, and the bran from 7.3 to 9.3%. These results are in good agreement with earlier studies of Bidwell (1918) and Bidwell et al. (1922). The bran as separated by Hubbard et al. after a tempering operation consisted of the cuticle, epidermis, hypoderm, and the major portion of the mesocarp. The innermost parts of the mesocarp, the micellar layer, and the aleurone remained with the endosperm fraction.

The composition of sorghum grain, as shown in Table 39, is similar to corn in many respects. This similarity extends to characteristics of the starch and protein, as well as some of the other components. The first column of ranges includes values typical of those obtained from analyses of commercial samples before the dominance of hybrids. The results in the second and third columns are more representative of those obtained on the commercial samples of the present day. In general, the change has resulted in a decrease in

Courtesy of Dr. S. A. Watson, Corn Products Co.

FIG. 29.   TRANSVERSE SECTION OF A
STEEPED SORGHUM KERNEL

TABLE 39

COMPOSITION OF SORGHUM GRAIN

| Component | Range | | Average[3] |
	Early[1]	Recent[2]	
Moisture	8–20	11–14	15.5
Protein	6.6–16.0	7.5–9.0	11.2
Fat	1.4–6.1	2.7–3.5	3.7
Ash	1.2–7.1	1.3–1.7	1.5
Reducing sugars[4]	0.4–2.5	...	1.8
Starch	60–77		74.1
Crude fiber	0.4–13.4	1.4–1.8	2.6
Tannin	0.003–0.17	...	0.1
Wax	0.2–0.5	...	0.3
Pentosans	1.8–4.9	...	2.5

[1] Miller (1958).
[2] Werler (1967).
[3] Watson (1967).
[4] As dextrose.

protein content and an increase in starch content. The averages for moisture, starch, protein, fat, ash, and fiber are averages for sorghum grain received at Corpus Christi, Texas, for use in starch manufacture during 1959 to 1962 (Watson 1967).

Carbohydrates

Carbohydrates other than starch are present only in small amounts. Unpublished data quoted by Watson (1967) show that both waxy and regular types average 1.20% total sugars composed of approximately 0.85% sucrose, 0.09% glucose, 0.09% D-fructose, and 0.11% raffinose. Sweet varieties will contain more of these sugars, a total of 2.8% being given by Watson.

Starch.—Starch granules from grain sorghum are very similar to those from corn, but the diameter may reach 35 μ as opposed to a top of about 30 μ for corn starch, and sorghum starch usually appears to have more of the large granules. Waxy sorghum starch granules are much the same but have a maximum diameter somewhat larger than regular sorghum starch granules.

Kofler gelatinization temperatures (initiation, midpoint, and completion, respectively) for sorghum starch are 154°, 164°, and 172°F, and for waxy sorghum starch are 154°, 159°, and 165°F (Schoch and Maywald 1967).

Deatherage *et al.* (1955) determined the amylose content of starch extracted from about 30 samples of sorghum grown in the United

TABLE 40

COMPONENT PARTS OF SORGHUM KERNELS AND PROXIMATE ANALYSIS[1]

	Proportion of Kernel[2]	Composition of Kernel Parts[3]			
		Starch	Protein	Fat	Ash
Whole kernel					
Mean	. . .	73.8	12.3	3.6	1.65
Range	. . .	82.3–75.1	11.5–13.2	3.2–3.9	1.57–1.68
Endosperm					
Mean	82.3	82.5	12.3	0.6	0.37
Range	80.0–84.6	81.3–83.0	11.2–13.0	0.4–0.8	0.30–0.44
Bran					
Mean	7.9	34.6[5]	6.7	4.9	2.02[5]
Range	7.3–9.3	. . .	5.2–7.6	3.7–6.0[6]	. . .
Germ					
Mean	9.8	13.4[5]	18.9	28.1	10.36[4]
Range	7.8–12.1	. . .	18.0–19.2	26.9–30.6	. . .

[1] Hubbard *et al.* 1950.
[2] As per cent of whole kernel, dry basis.
[3] Per cent, dry basis.
[4] Includes wax, 0.29–0.44%, mean 0.32% db.
[5] Composite.
[6] Includes five varieties.

States and about 175 samples of sorghum obtained from foreign sources. The amylose content of the starch from nonwaxy types ranged from 21 to 28% with a mean of 25%. There was no appreciable difference in amylose content between the foreign and the US samples. The waxy samples had apparent amylose contents of 1 or 2%.

Lipids

Fatty acid composition of a typical crude and dewaxed oil was reported by Kummerow (1946) to be myristic 0.2%, palmitic 8.3%, stearic 5.8%, hexedecenoic 0.1%, oleic 36.2%, and linoleic 49.4%. Further details of crude oils, as reported by two investigators, are shown in Table 41.

TABLE 41

CHARACTERISTICS OF THE FAT OF SORGHUM

	Baird (1910)	Kumerow (1946)
Specific gravity	0.94 gm/ml	...
Melting point	111.6°F	...
Iodine number	109.7	119.0
Saponification number	249.1	181.0
Reichert-Meissl number	6.10	...
Acetyl number	42.2	...
Refractive index at 77°F	...	1.4718

The petroleum ether extract of sorghum bran is greater than that of some other cereal brans, such as corn, and contains mostly wax which is semisolid at room temperature. This material can be processed to give a hard wax (Kummerow 1946).

Vitamins

Most varieties of sorghum contain no vitamin A activity. Typical contents of some of the vitamins present in sorghum are shown

TABLE 42

THE VITAMIN CONTENT OF SORGHUM AND ITS COMPONENT PARTS

Vitamin	Whole Kernel[1]	Proportion of the Total Found in Components[2] (%)		
		Endosperm	Germ	Bran
Niacin	51.4	75.6	17.1	7.3
Pantothenic acid	7.0	64.8	28.1	7.1
Riboflavin	1.11	50.5	27.7	21.8
Biotin	0.30	52.7	31.6	15.7
Pyridoxine	6.4	75.8	16.3	8.0

[1] Anon. 1958. Values in micrograms per gram.
[2] Hubbard et al. 1950.

in Table 42. The average distribution of these vitamins in five varieties (taken from another reference) is also shown in the table and indicates the presence of significant amounts in the endosperm and germ, while the bran is poorly provided with these nutrients. An earlier report, by Tanner *et al.* (1947), based on analyses of a larger number of samples, showed about twice the range of vitamin content reported by Hubbard *et al.* (1950).

UTILIZATION AND QUALITY

The principal use of sorghum grain in the United States is for animal feed, and this use accounts for about 90% of the sorghum grain produced. The grain is approximately equal to corn as a feed grain for most classes of livestock, although the differences in contents of protein, oil, and vitamin A should be considered. It may be used directly as a feed grain, or it may be processed by the mixed-feed industry into commercial feed preparations.

TABLE 43

ELEMENTAL ANALYSIS OF THE ASH OF REDBINE-66 SORGHUM GRAIN GROWN IN MOORE COUNTY, TEXAS, IN 1951, PER CENT OF GRAIN AS RECEIVED[1]

Element	%[2]
Phosphorus	0.4
Potassium	0.3
Magnesium	0.2
Calcium	0.02
Iron	0.02
Manganese	0.02
Zinc	0.008
Silicon	0.004
Boron	0.001
Lead	0.001
Aluminum	0.0004
Molybdenum	0.0001
Copper	0.00008
Nickel	0.00008
Sodium	0.00008
Chromium	0.00004
Tin	0.00004
Titanium	0.00004

[1] From McNeill (1952).
[2] By semiquantitative spectrographic analysis.

Little attempt is made to segregate sorghum grain on the basis of its components. The basic document controlling quality designations of grain sorghum destined for use as feed is the Official Grain Standards of the United States (Anon. 1964). Under these standards, grain sorghum is grain which, before the removal of dockage, consists of 50% or more of whole kernels of grain sorghum and not more than 10% of other grains listed in the Standards. Grain sorghum is divided into the classes of yellow, white, brown, and mixed, based on the color of the seedcoats. Ten per cent of kernels of colors other than the named color is permitted in the first three classes.

Table 44 lists the grades and grade requirements for the classes Yellow Grain Sorghum, White Grain Sorghum, Brown Grain Sorghum, and Mixed Grain Sorghum. The test weight per bushel (column 2) is based on the Winchester bushel.

TABLE 44

UNITED STATES GRADES, GRADE REQUIREMENTS AND GRADE DESIGNATIONS FOR GRAIN SORGHUM[1]

| | | | Per Cent Maximum Limits of | | |
| | Minimum Test Weight per Bushel (Lb) | Mois-ture | Damaged Kernels | | Broken Kernels, Foreign Material, and Other Grains |
Grade			Total	Heat Damaged	
1	57	13.0	2.0	0.2	4.0
2	55	14.0	5.0	0.5	8.0
3[2]	53	15.0	10.0	1.0	12.0
4	51	18.0	15.0	3.0	15.0
Sample grade[3]	...	...	...	...	...

[1] Anon. 1964.
[2] Grain sorghum which is distinctly discolored shall not be graded higher than No. 3.
[3] Sample grade shall be grain sorghum which does not meet the requirements of any of the grades from No. 1 to No. 4, inclusive; or which contains stones; or which is musty, or sour, or heating; or which is badly weathered; or which has any commercially objectionable foreign odor except of smut; or which is otherwise of distinctly low quality.

There are two special grades not listed in the table. These are for Smutty Grain Sorghum and Weevily Grain Sorghum.

Although the flavor of the unprocessed grain and flour is too pronounced to allow their ready acceptance by persons not accustomed to consuming them from an early age, the economic advantage over other cereals (sorghum is usually even cheaper than corn) leads to continuing efforts to apply the flour as an ingredient of human foods. Various processing methods have been developed to reduce the characteristic flavor. These generally rely on some form of steam or dry heat treatment. Small amounts of such flours have been used in soda crackers, snack foods, etc. The quality features of greatest importance are freedom from off-flavors and light color.

In many parts of Asia and Africa, sorghum is the most important food grain and makes up a large part of the diet for the majority of the population. In these areas, some form of the grain is eaten at each meal.

A fair amount of waxy sorghum starch is used in foods. The

starch is separated by a wet-milling process similar to that used in making cornstarch, and this procedure removes most, if not all, of the undesirable flavor. It is important that grain for such processing be free of off-flavors, light in color, and with minimal amylase activity. The starch fraction should, of course, be as near as possible to 100% amylopectin.

Although the emphasis of this book is on the food and feed applications of the cereal grains, it might be instructive to consider some of the quality factors affecting sorghum suitability for industrial use. Most of these uses take advantage of sorghum flour as a low-cost source of starch. High quality for such uses would require high starch content, low fiber and protein content, and, usually, light color. Grits used in dog food must conform to a specification of 8.5% minimum protein content for certain customers. Since this requirement is difficult to meet with mill-run grain, the quality factor of protein content is important in this case.

Although sorghum grits are not currently being used by brewers, this is an outlet of large potential volume, and, at one time, several million pounds of grits were used for this purpose. Beer made with sorghum grits takes on a bitter flavor due to the anthocyanogens present in the grain. Breeding of varieties to eliminate substantially all of these substances, or specialized milling processes, might lead to a substantial use of sorghum grits in brewing. Light-colored grain with a minimum of color staining would also be required for this application.

Sorghum flours are used as strength additives in building materials made of wood fiber, mineral wool, or gypsum. Crude flours are used as flocculating agents in aluminum ore refining. Gelatinized and ungelatinized sorghum flours are used as binders in charcoal briquettes. No information is available on special quality requirements for these uses.

The preceding discussion on industrial uses of sorghum grain is based in large part on the paper of Werler (1967).

BIBLIOGRAPHY

ANDERSON, E., and MARTIN, J. H. 1949. World production and consumption of millet and sorghum. Econ. Botany *3*, 265.

ANON. 1958. Unpublished data. Northern Regional Res. Lab., US Dept. Agr., Peoria, Ill.

ANON. 1964. Official Grain Standards of the United States. US Dept. Agr. Publ. SRA-AMS *177*.

ANON. 1966. Annual Summary of Crop Production. US Dept. Agr., Washington, D. C.

BAIRD, R. O.   1910.   The chemistry of Kafir corn kernel.   Okla. Agr. Expt. Sta. Bull. *89*.

BALL, C. R.   1910.   The history and distribution of sorghum.   US Dept. Agr. Bur. Plant Industry Bull. *175*.

BAUMGARTEN, W., MATHER, A. N., and STONE, L.   1946.   Essential amino acid composition of feed materials.   Cereal Chem. *23*, 135–155.

BIDWELL, G. L.   1918.   The physical and chemical study of the kafir kernel.   US Dept. Agr. Bull. *634*, 6.

BIDWELL, G. L., BOPST, L. E., and BOWLING, J. D.   1922.   A physical and chemical study of milo and feterita kernels.   US Dept. Agr. Bull. *1129*, 1–8.

DEATHERAGE, W. L., MACMASTERS, M. M., and RIST, C. E.   1955. A partial survey of amylose content in starch from domestic and foreign varieties of corn, wheat, and sorghum and from some other starch-bearing plants.   Trans. Am. Assoc. Cereal Chemists *13*, 31–42.

HUBBARD, J. E., HALL, H. H., AND EARLE, F. R.   1950.   Composition of the component parts of the sorghum kernel.   Cereal Chem. *27*, 415–420.

KARPER, R. E., and QUINBY, J. R.   1947.   Sorghum—its production, utilization and breeding.   Econ. Botany *1*, 355–371.

KRAMER, N. W.   1957.   Unpublished data on optimum plant populations of sorghum for different productivity levels.

KRAMER, N. W.   1958.   Hybrid sorghums for grain and forage.   Proc. Third Ann. Farm Seed Ind.-Res. Conf.   American Seed Trade Assoc., Chicago, Ill.

KUMMEROW, F. A.   1946.   The composition of sorghum grain oil, andropogon sorghum var.   vulgaris.   Oil and Soap *23*, 167–170.

MARTIN, J. H., and LEONARD, W. H.   1949.   Principles of Field Crop Production.   Macmillan Co., New York.

MARTIN, J. H., and MACMASTERS, M. M.   1951.   Industrial uses for grain sorghum.   US Dept. Agr. Yearbook Agr. *1951*, 349–352.

McNEILL, J. R.   1952.   Unpublished data.

MILLER, D. F.   1958.   Composition of Cereal Grains and Forages.   Natl. Acad. Sci. Natl. Res. Council Publ. *585*.

MILLER, E. C.   1916.   Comparative study of the root systems and leaf areas of corn and the sorghums.   J. Agr. Res. *6*, 311–333.

MURTY, B. R., ARUNACHALAM, V., and SAXENA, M. B. L.   1967.   Classification and catalog of a world collection of sorghum.   Indian J. Genetics Plant Breeding *27*, 1–312.

NELSON, G. H., TALLEY, L. E., and ARONOVSKY, S. I.   1950.   Chemical composition of grain and seed hulls, nut shells, and fruit pits.   Trans. Am. Assoc. Cereal Chemists *8*, 58–68.

QUINBY, J. R., KRAMER, N. W., STEPHENS, J. C., LAHR, K. A., and KARPER, R. E.   1958.   Grain sorghum production in Texas.   Texas Agr. Expt. Sta. Bull. *912*.

ROONEY, L. W., and CLARK, L. E.   1968.   The chemistry and processing of sorghum grain.   Cereal Sci. Today *13*, 258–287.

SCHOCH, T. J., and MAYWALD, E. C.   1967.   Industrial microscopy of starches.   *In* Starch: Chemistry and Technology, Vol. 2.   R. L.

WHISTLER, and E. F. PASCHALL (Editors). Academic Press, New York.

SWANSON, N. P., and THAXTON, E. L., JR. 1957. Requirements for grain sorghum irrigation on the High Plains. Texas Agr. Expt. Sta. Bull. *846*.

TANNER, F. W., JR., PFEIFFER, S. E., and CURTIS, J. J. 1947. B-complex vitamins in grain sorghums. Cereal Chem. *24*, 268–274.

VINALL, H. N., STEPHENS, J. C., and MARTIN, J. H. 1936. Identification, history, and distribution of common sorghum varieties. US Dept. Agr. Tech. Bull. *506*.

WAGNER, W., TSCHANG, T. S., LIU, T. H., and SHIA, A. Y. 1913. The nutrient uptake and fertilizer requirements of sorghum. Berichte aus der Deutsch-Chinesischen Hochschule. Tsingtau, China.

WATSON, S. A. 1967. Manufacture of corn and milo starches. *In* Starch: Chemistry and Technology, Vol. 2. R. L. WHISTLER and E. F. PASCHALL (Editors). Academic Press, New York.

WATSON, S. A., and HIRATA, Y. 1955. The wet milling properties of grain sorghums. Agron. J. *47*, 11–15.

WERLER, P. F. 1967. Manufacture and usage of sorghum flour. Proc. 8th Ann. Symp. Central States Sect. Am. Assoc. Cereal Chemists. St. Louis, Mo. Feb. 17–18.

CHAPTER 7

Samuel A. Matz

H. M. Beachell[1]

Rice

INTRODUCTION

Rice, one of the oldest and most important food crops, is the staple food of over half the world's population. World rice acreage has approached 230 million acres during the last few years. Annual world production of rough rice probably lies between 350 and 400 billion pounds.

About 92% of the world rice crop is produced in Asia and adjacent island areas of dense population. An even higher percentage is consumed in these areas so that most Asiatic countries are deficient in rice and only Burma and Thailand, have an exportable surplus.

Rice probably originated somewhere in Southeast Asia, where it has been grown for many centuries. The earliest record of rice production in China dates back to about 2800 BC (Ghose *et al.* 1956) and in India back to 1000 BC. Later, rice culture spread westward to the Middle East, Africa, and Europe. It was cultivated in the Euphrates Valley in 400 BC, was mentioned by the Greek poet Sophocles in the Tragedies in 495 BC, and was brought to Southern Europe in Mediaeval times by the Saracens. It has long been an important crop and food in Spain, Italy, and Portugal.

Rice culture in the United States began about 1685 in South Carolina and later spread to North Carolina, Georgia, Alabama, Mississippi, and Florida (Jones 1936). As late as 1849, South Carolina, North Carolina, and Georgia produced about 90% of the US rice crop.

Following the Civil War, rice production in the South Atlantic states decreased rapidly, followed by an increase in acreage along the lower Mississippi River in Louisiana. By 1900 rice production had spread to the coastal prairies of Louisiana and Texas and by 1905 rice was being grown in the Grand Prairie of southeastern Arkansas. By 1909 Louisiana and Texas were producing about 90% of the US rice crop. In 1912 the first commercial acreage of rice was grown in California.

Today, the five major rice-producing states are Texas, Louisiana,

---

[1] H. M. BEACHELL is affiliated with the Rockefeller Foundation and located at the International Rice Research Institute, Manila, The Philippines.

Arkansas, California, and Mississippi. Rice has also been grown commercially in Missouri, along the Mississippi River as far north as Palmyra. The average farm value of the US rice crop for the 5-yr period 1959–1965 was about $300,000,000. It is the major cash crop in most of the counties and parishes where grown.

BOTANY OF THE RICE PLANT

Cultivated rice (*Oryza sativa*) belongs to the Gramineae or grass family and the tribe Oryzeae. Unlike the other cereal crops, rice is a semiaquatic plant which thrives under flooded soil conditions. The chromosome number of *Oryza sativa* is 24 (2N). Among the wild species, both 24 and 48 chromosome types are reported.

Over 60 species of *Oryza* have been reported, but only 23 valid species are generally recognized, according to Chatterjee (1948). Most of the wild species are found in Southeast Asia and Africa, but several species have been reported from the Western Hemisphere. A list of the more commonly recognized species of *Oryza* are shown in Table 45. Chang (1968) describes a species *Oryza*

TABLE 45

THE SPECIES OF ORYZA MOST COMMONLY RECOGNIZED, SHOWING THEIR CHROMOSOME NUMBER AND AREA OF THE WORLD WHERE FOUND[1]

Species of Oryza	Number of Chromosomes (2N)	Country of Origin
Oryza sativa (Cultivated rice)	24	Asia
Oryza glaberrima (Cultivated in Africa)	24	Africa
Oryza sativa var. *fatua* (*spontanea*)	24	Asia
Oryza alta	48	America
Oryza australiensis	24	Australia
Oryza brachyantha	24	Africa
Oryza breviligulata	24	Africa
Oryza coarctata	48	Asia
Oryza cubensis—sometimes classed under *Oryza perennis*	24	America
Oryza eichingeri	48	Africa
Oryza grandiglumis	24	America
Oryza granulata	24	Asia
Oryza latifolia	48	America
Oryza meyeriana	24	Asia
Oryza minuta	48	Asia
Oryza officinalis	24	Asia
Oryza perennis	24	Asia, Africa, America
Oryza perrieri	24	Malagasy
Oryza punctata	24, 48	Africa
Oryza ridleyi	48	Asia
Oryza schlechteri	...	New Guinea
Oryza stapfii—sometimes classed under *Oryza sylvestris*	48	Africa
Oryza subulata	24	America
Oryza tisseranti	...	Africa

[1] Formulated from information reported by Chatterjee (1948), Jones and Longley (1941), Ramiah and Rao (1953), and Sampath and Rao (1951).

rufipogon (2N = 24) which includes *O. perennis, O. sativa f. spontanea,* and *O. cubensis.*

Cultivated rice does not cross freely with most wild species, except that *sativa* and *glaberrima* intercross freely (Ramiah and Roa 1953). A number of interspecific crosses have been reported by Ramiah and Ghose (1951) and Sampath and Rao (1951), but only limited information on the subject is available. Further cytogenetic studies of interspecific hybrids are needed before the relationship between cultivated rice and the various wild species can be established. Ramiah and Ghose (1951) reported that useful forms somewhat resistant to drought were obtained from a cross between *Oryza sativa* and *Oryza perennis.*

Ramiah and Rao (1953) state that over 8,000 botanically different varieties of rice are in existence in the world and that more than 4,000 varieties have been identified in India. Jones *et al.* (1953) report that over 8,000 varieties and selections were introduced into the United States for testing purposes during the 50-yr period 1903 to 1952. At the present time over 10,000 varieties are maintained in the World Rice Collection of the International Rice Research Institute (Anon. 1966).

The varieties of *Oryza sativa* are usually divided into two subspecies, *japonica* and *indica.* Ghose *et al.* (1956) based their classification on morphological differences, and response to temperature and length of day. All of the *japonica* types have essentially the same grain and panicle type, according to Ramiah and Ghose (1951). Crosses between *indica* and *japonica* varieties show varying degrees of sterility. Jones and Longley (1941) concluded that *japonica* and *indica* varieties probably had a common origin many centuries ago and that gene mutations and chromosomal rearrangement are responsible for this incompatibility. In the United States, the *japonica* and *indica* subspecies have been intercrossed and several important commercial varieties developed. In Asia, *japonica-indica* crosses have frequently produced plants with a high degree of sterility. An explanation for the difference in results, based on recent studies, is that fertility is largely determined by the genetic structure of the *indica* parent. All (or most) crosses made with certain *indica* varieties will yield fertile offspring, while other varieties will cause sterility regardless of the *japonica* variety used as the other parent. At IRRI, *japonica* and *indica* varieties have been intercrossed and partial sterility of the F_1 generation plants is a common observation. However, no difficulty was experienced in obtaining fertile lines in later generations. A more complete description of the *japonica* and *indica* varieties will be undertaken in discussing the varieties grown in the United States.

The rice plant has a branched panicle type inflorescence. The spikelets are borne on the panicle branches, and each spikelet contains a single ovule and six anthers. The ovule when fertilized forms the rice grain, which is covered by loose outer hulls called the lemma and palea. The hulls remain intact when threshed and are removed as a first step in the milling operation. At the base of the lemma and palea are two outer sterile glumes which remain attached to the grain when threshed and are usually greatly reduced in size.

The rice florets usually bloom after the panicles emerge from the leaf sheath. Blooming usually occurs at about the same time of day in each region where grown, but the exact time depends upon the daily fluctuations in temperature and humidity. In the United States this occurs between the hours of 8:00 AM and 4:00 PM with most of the blooming occurring between 10:00 AM and 12:00 noon (Adair 1934 and Laude and Stansel 1927). Blooming is characterized by the opening of the lemma and palea. Simultaneous with the opening of the glumes is a rapid elongation of the styles bearing the anthers. Pollination takes place as the anthers dehisce, liberating the pollen grains which fall on the feathery stigma attached to the ovule. The glumes usually remain open about one hour.

Rice is a self-fertilized crop, but some natural crossing occurs. Natural crossing has been reported to vary from a fraction of 1% to over 3% (Beachell *et al.* 1938).

PRODUCTION AND TRADE STATISTICS

Rice production is widely distributed over the world in spite of the heavy concentration of production in Southeast Asia. It is grown as far north as northern Japan (Hokkaido), and as far south as Argentina and Chile. A list of the most important rice-growing countries and statistics related to their production and consumption of this grain are given in Table 46.

Average yields vary from over 4,500 lb per acre in Italy to slightly over 1,000 lb per acre in Cambodia and Laos. In general, the tropical countries produce low yields in comparison with nations in the temperate regions. The low yields in the tropics are due largely to the low level of cultural inputs there and to the fact that much of the crop is dependent upon the monsoons for irrigation water, which, as a result, may or may not be adequate. Drainage

TABLE 46

ACREAGE, PRODUCTION, AND CONSUMPTION OF RICE IN LEADING
RICE-GROWING COUNTRIES[1]

Country	Acreage[2]	Produc- tion[3]	Average Yield[4]	Per Capita Produc- tion[5]	Per Capita Dis- appearance[6]
Brazil	9,800	13,887	1,417	171	111
United States	1,786	7,311	4,094	37.4	9.4
Italy	296	1,361	4,598	26.4	14.4
Burma	12,500	17,860	1,429	723	342
Cambodia	5,535	5,828	1,053	1,015	481
Ceylon	1,088	1,817	1,670	162	234
China (Taiwan)	1,884	6,590	3,498	531	322
India	89,148	128,212	1,438	272	180
Indonesia	18,500	30,860	1,668	309	222
Japan	8,055	34,681	4,306	354	239
Korea, South	2,954	8,770	2,969	309	200
Laos	1,643[7]	1,698[7]	1,033[7]	566	399
Malaysia	995	2,200	2,211	241	267
Pakistan	26,155	39,232	1,500	381	247
Philippines	7,919	8,822	1,114	273	198
Thailand	14,814	21,219	1,433	693	312
Vietnam, South	6,313	11,091	1,757	688	441
Malagasy Republic	1,853	2,932	1,582	474	300
United Arab Republic	999	4,488	4,493	169	66

[1] Based on data from Agricultural Statistics 1966. Data for 1964 where available.
[2] Thousands of acres harvested.
[3] Millions of pounds of rough rice.
[4] Pounds of rough rice per acre harvested.
[5] Pounds of rough rice per person.
[6] Pounds per person of milled rice equivalent.
[7] Average of 1955 through 1959 data.

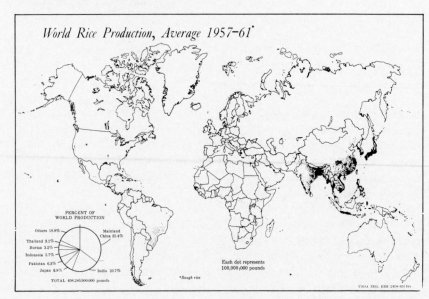

Courtesy of US Dept. Agr.

Fig. 30.   Distribution of World Rice Production

may also be poor or nonexistent in these countries, and the same lands are cropped to rice year after year, with little fertilizer being added. Frequently, lands poorly adapted for rice are cultivated in an effort to meet the continually increasing needs for food.

In Japan, comparatively high yields are obtained (4,306 lb per acre). This is the result of many years devoted to research in the development and use of adapted varieties and good cultural practices. Controlled irrigation, soil-building crops, and commercial fertilizers also contribute to the higher yields obtained in Japan. The extremely high yields in Italy (and in countries such as Spain and Portugal, which are not included in this table) are due to the use of soil-building rotation systems, the application of large amounts of commercial fertilizers, and the temperate climate, which results in fewer disease and insect problems.

The *per capita* production can be instructive, but the data need to be related to the actual work force involved in rice production. In countries such as Cambodia, a large percentage of the population is engaged in rice-farming, resulting in high *per capita* yields, while Italy and the United States have relatively small numbers of people working on rice farms, with the result that the *per capita* production figures for the latter countries are very low.

The disappearance figures should be understood as including losses through nonfood routes. For example, destruction by rodents and insects may account for some of the very high consumptions listed in the table. Military activities may cause considerable losses in some countries. The estimate for the United States includes such uses as brewers' adjuncts, and may be compared with the US Dept. of Agr. estimate of 7.0 lb consumed directly as milled rice. For *ad libitum* consumption in areas where rice makes up most of the diet for the average person, an actual *per capita* use as food approximating 300 lb per year can be surmised. The average adult consumption would probably be half again as much, leading to a rough estimate of 1.25 lb of milled rice per day as the amount consumed by an adult who can eat as much of the grain as he wants. This would provide slightly more than 2,000 cal.

Average acre yields produced in the United States from 1899 to 1967 are shown by 2- and 5-yr averages in Table 47. It can be seen that the average yield for the 2-yr period 1899 and 1900 was only 1,207 lb per acre, but by the period 1906 to 1910 it had increased to 1,642 lb. Little further change in average yields occurred until 1926 to 1930, when the average increased to 2,000 lb per acre. This increase can be attributed mostly to the develop-

AVERAGE YIELDS OF ROUGH RICE IN THE UNITED STATES DURING THE PERIOD
1899 to 1967[1]

Years	Annual Yield in Lb per Acre
1899–1900	1,207
1901–1905	1,442
1906–1910	1,642
1911–1915	1,599
1916–1920	1,791
1921–1925	1,754
1926–1930	2,000
1931–1935	2,137
1936–1940	2,257
1941–1945	2,005
1946–1950	2,161
1951–1955	2,523
1956–1960	3,265
1960–1965	3,898
1967	4,531

[1] Anon. 1968A.

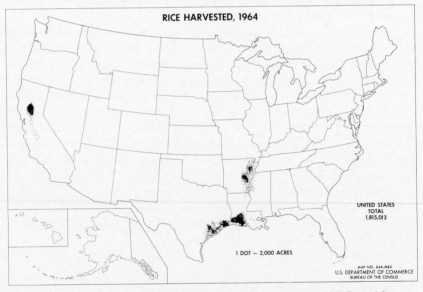

Courtesy of US Dept. of Commerce

FIG. 31. RICE HARVESTED IN THE UNITED STATES

ment of tractors and improved implements that could be operated successfully under rice land conditions. Earlier practice had been to use mules and unadapted tractors for power. Following this rise, there was another period of relative stability which lasted until about 1951 to 1955, when the 5 yr average yield increased to 2,523. A slight drop which occurred from 1941 to 1950 was due to the intense cropping of land, and perhaps a labor shortage, during and immediately after World War II. In 1957, a yield of 3,068 lb per acre was obtained, and this stood as an all-time high for several years. The increases in yield during the 1951 to 1955 period can be attributed in large part to the increased use of commercial fertilizers, the use of herbicides for controlling weeds, improved cultural practices, and better varieties. The 1956 and 1957 rise in yield was brought about in part through Federal acreage controls. Farmers no doubt planted their best lands in rice, abandoning the poorer acreage or using it for other crops, and practiced better farming methods.

The extremely high yields in recent years are due to a continuing improvement in equipment and land management, and use of very early varieties which permits stubble cropping.

Average yields produced in the different states vary considerably. California has consistently produced higher average yields than the other states, as shown in Table 48, although the spread has lessened in recent years. The higher yields in California are partly due to soil and climate and partly to the varieties grown.

TABLE 48

ACREAGE AND YIELDS IN THE MAJOR RICE-GROWING STATES DURING RECENT YEARS[1]

	1946–1950	1951–1955	1957–1961	1962–1965	1967
Arkansas					
Acreage[2]	359	498	364	429	479
Average yield[3]	2,200	2,370	3,307	4,153	4,669
Louisiana					
Acreage[2]	594	602	435	511	566
Average yield[3]	1,768	2,210	2,797	3,332	3,904
Mississippi					
Acreage[2]	6[4]	51	40	49	55
Average yield[3]	2,700[4]	2,515	3,010	3,657	4,224
Texas					
Acreage[2]	483	562	394	461	508
Average yield[3]	2,077	2,650	3,081	4,104	4,938
California					
Acreage[2]	255	372	268	325	361
Average yield[3]	3,216	3,185	4,604	4,827	4,805

[1] Computed from Agricultural Statistics, US Dept. Agr. and Anon. 1968A.
[2] In thousands of acres harvested.
[3] In pounds of rough rice per acre.
[4] Two-year average, 1949 and 1950.

The relatively low yields in Louisiana are due to the fact that the lands used for rice production have been heavily cropped to this grain for many years. From the early 1900's until fairly recently, much of the rice land of Louisiana produced a rice crop every other year. The same is true of some of the lands in the eastern part of the Texas rice area. Broadleaf weeds, annual grasses, and sedges have heavily infested the fields and have been at least partly re-

TABLE 49

EXPORT AND IMPORT ACTIVITY OF COUNTRIES PROMINENT IN
INTERNATIONAL RICE TRADING[1]

Country or Region	Average Annual Exports for the Years[2]		Average Annual Imports for the Years[2]	
	1956–1960	1963–1964	1956–1960	1963–1964
North America				
Canada	2.9	...	98.2	102.5
Cuba	...	...	368.4	419.0
United States	1,787.3	2,771.6	27.7	1.2
Europe				
Belgium and Luxembourg	56.1	16.5	154.7	100.3
France	...	...	30.4	28.4
Italy	424.2	237.9	2.0	7.3
Netherlands	65.9	40.5	160.7	139.3
Spain	136.3	121.8	...	...
West Germany	10.8	29.4	278.9	335.0
United Kingdom	1.7	...	192.3	244.0
Asia				
Arabian Peninsula	45.3	41.0	206.1	511.4
Burma	3,736.3	3,415.0	...	...
Ceylon	4.5	0.2	1,150.1	1,169.5
China, Mainland	2,544.3	1,515.7	...	...
China, Taiwan	274.1	272.3	7.9	2.2
Hong Kong	56.6	84.2	784.0	905.8
Cambodia	442.7	926.4	...	...
South Vietnam	388.9	409.2	39.5	...
Japan	...	...	1,971.0	2,382.6
South Korea	13.6	19.7	102.6	129.2
Malaysia and Singapore	213.4	204.4	1,457.6	1,382.7
Pakistan	101.1	304.6	811.8	392.8
Philippines	0.8	...	157.6	613.0
Ryuku Islands	...	...	140.8	180.0
Thailand	2,774.4	3,683.8	...	...
Africa				
Mauritius	...	...	130.7	156.2
Senegal-Sudan-Mauritius	...	...	209.5	314.4
United Arab Republic	543.3	998.6	25.4	...
Oceania				
Australia	101.0	167.4	...	...

[1] Agricultural Statistics 1966.
[2] Quantities are in millions of pounds milled rice equivalent. No entry means less than 100,000 or inadequate data available.

sponsible for the low yields.   Improved cultural practices such as land leveling, chemical weed control, and water seeding have helped to raise the average yields on these lands.   Stubble or ratoon cropping, particularly in Texas, has been responsible for considerable increases in yields.   By using varieties which mature in about 100 days, a second crop can be obtained from the ratoons.   As a result, the 1967 Texas yield of 4,938 lb per acre exceeded the California yield.

Exports and Imports

Total world exports and imports of milled rice are shown in Table 49.   It is obvious that only a small percentage of the total world rice crop moves in international trade.   In general, the countries that grow a great deal of rice also consume large quantities. Perhaps seven per cent of the world crop is exported from the country of origin.

The leading exporting countries are Burma, Thailand, and the United States.   These countries are presently accounting for over 60% of the export trade in rice.   The United States, a country producing about 2% of the world crop, accounts for roughly 20% of the world rice export.   The United States annually exports over half of its total production.

The leading importing countries are Ceylon, India, Indonesia, Japan, and Malaysia (including Singapore).   Outside of Asia, the most prominent importers are Cuba, Western Germany, the United Kingdom, and the Senegal-Sudan-Mauritania region.

THE CULTURE OF RICE

Climate

Rice can be successfully grown under a wide range of climatic conditions.   Most of the world rice crop is grown between the Equator and 40°N latitude, but rice culture extends from 45°N to 40°S latitude.   The crop is commonly associated with tropical regions, but it is also widely grown in temperate regions.   The many types of rice which have evolved through the centuries of extensive rice culture vary widely in their range of adaptability. Rice generally requires a minimum growing season of 4 to 5 months, during which the mean temperature must average 70°F or above. The crop is grown at altitudes from sea level to over 5,000 ft (Efferson 1952 and Ghose et al. 1956).

In tropical humid regions, tall rank-growing long-season varieties are usually grown. In tropical Asian countries varieties requiring a growing season of over 180 days are common. These tropical varieties grow tall, tiller profusely, and produce rank vegetative growth on soils that are annually cropped to rice. In low-lying flood plains deep-water rices, commonly called floating rices, are grown. These rices elongate as much as 12 in. a day as the waters rise and may reach a height of 12 to 20 ft.

In temperate regions short-season varieties maturing in 85 to 145 days are grown. They are usually hardy, somewhat short-stature types that can withstand wide variations in daily temperatures and produce satisfactory yields even when grown in cold irrigation water.

The IRRI reported that yield of grain was correlated with the amount of sunlight (measured in cal/cm^2 /day) received over the last 4 to 6 weeks before harvest. Different varieties differed in the efficiency of their use of radiant energy.

Soil Types

Rice is grown on a wide range of soil types. The lack of adequate irrigation water is more apt to be a limiting factor than is soil type. The soil pH can vary from 4.0 to 8.0 (Ghose *et al.* 1956). Soils with impervious subsoils capable of holding flood water or level prairies that can be readily flooded make ideal rice lands.

Deep soils tend to give higher yields than soils that have an impenetrable layer at a shallow depth. This effect must be balanced against the advantage of an impervious subsoil for retaining the irrigation water. Experiments at IRRI showed that, as the depth of soil was increased to about 40 cm, the grain yield increased in proportion. Beyond 40 cm there was some yield improvement, but it was relatively slight. The soil in question was a fertile clay soil. Fertilizer nitrogen was much more effective on shallow than on deep soils. The disadvantages of shallow soils could be largely eliminated by suitable fertilizer application.

The lands used for commercial rice production in the United States are usually comparatively level and have reasonably good surface drainage. Medium or heavy clays, clay loams, silt loams, or fine sandy loams with slowly permeable subsoils are preferred. Such soils can be flood-irrigated more efficiently and will support heavy mechanized equipment under a wider range of soil moisture conditions than lighter soils. Soils of this type occur in the coastal prairie region of southwest Louisiana and southeastern Texas (Walker and Miears 1957). They also occur in the Grand Prairie

Courtesy of Louisiana Rice Experiment Station

FIG. 32. PLOWING LAND IN PREPARATION FOR SEEDING

and other areas in eastern Arkansas, elsewhere in the south central states and in the interior valleys of California. The rice plant has a high alkali tolerance and is sometimes grown for reclaiming such lands (Efferson 1952).

Water Requirements and Irrigation

The commercial rice crop of the United States is grown as an irrigated crop, and fields are kept flooded throughout the greater part of the growing season. The rice plant requires abundant soil moisture and usually thrives under continually flooded soil conditions. This requires an abundant supply of readily available irrigation water throughout the growing season. Flooding of fields provides a means of controlling many serious weed pests and assures adequate moisture at all times for optimum growth and development of the rice plant.

The total water requirements of a rice crop vary greatly depending upon the type of soil, climatic conditions, and the efficiency of irrigation systems. Davis (1950) estimates that from 4 to 10 acre-ft of water are required to produce a rice crop in California. Jones *et al.* (1952) report that 40 to 45 in. of water are required in Texas and Louisiana and about 33 in. are required in Arkansas. The water requirements supplied by rainfall during the growing season amount to about 15 in. in Texas and Louisiana, about 11 in. in Arkansas, and a negligible amount in California.

In the southern states the land may be drained one or more times during the growing season, but in California the land is rarely drained from the time the seed is sown until harvest. The rice plant also grows well under nonflooded or upland conditions. In parts of South America and Asia upland rice is an important crop. Efferson (1952) states that much of the South American rice crop is grown under nonflooded or upland conditions. Upland rice is grown throughout Asia, but it constitutes only a small percentage of the total acreage. Short-season varieties usually maturing in 100 to 125 days are used. The land is cleared and the rice grown during seasons of frequent rainfall. After several crops, yields usually drop off due to reduced fertility and the intrusion of weeds. The area is then abandoned and another area cleared. Yields are low, but little effort is required to produce a crop. Prior to World War II, small acreages of upland rice were grown in Alabama, Florida, Georgia, Mississippi, and South Carolina, but the production of this type was less than 1% of the total US rice crop. Upland rice was grown as a cultivated crop sown in rows $1^1/_2$ to 3 ft apart (Jones 1943). Today this culture is practically nonexistent.

In the United States, irrigation water is obtained from bayous and streams, and from deep wells. According to Adair and Engler (1955), over 40% of the 1953 rice crop in this country was irrigated from wells. They estimated the percentage of the crop irrigated from wells to be about 90% in Arkansas, 40% in Louisiana, 20% in Texas, and 10% in California.

In Texas and Louisiana, surface water obtained from bayous and streams is pumped into canals which usually flow by gravity into laterals leading into the rice fields. The size of the irrigation systems varies from a few thousand acres to over 60,000 acres, and the water is lifted as much as 50 ft into the canals.

In some sections of Arkansas, Louisiana, and Texas, small pumping systems are installed to elevate surface water from streams and ditches into reservoirs during periods of abundant surface water supply. The water from the reservoir is either pumped or allowed to flow by gravity to the rice fields as it is needed.

In California, most of the water used for rice irrigation is obtained from large streams. Since the water may be rather cool, warming basins are frequently used to raise the water temperature before irrigating the fields. Hardy *japonica*-type varieties are grown in California to help overcome the cool water conditions.

Range of Cultural Practices and Rotation Systems

In tropical Asia, essentially all rice is produced by intense hand methods of production. The land is usually plowed and puddled under flooded conditions, using oxen. In many areas, rainfall is depended upon for irrigating the crop. Harvesting, threshing, and milling are hand operations (see Fig. 42). As a result, there is an average labor requirement of over 200 man-days to produce an acre of rice.

In the United States, highly mechanized production methods are used to cut the labor requirement to less than two man-days for each acre of rice. Much the same general type of equipment is used as for producing other small grain crops, except that the special requirements resulting from flood irrigation must be met. Flood irrigation is accomplished by smoothing the land and then building contour levees or dikes which are located by survey and usually placed at elevation differences of $^2/_{10}$ ft in the Southern States and $^3/_{10}$ ft in California.

In tropical Asian countries lands are cropped continuously to rice. When water is available two crops per year are grown. This has been practiced for centuries and yields are low. However, IRRI has demonstrated that yields of 4,000 to 6,000 kg/ha can be produced on these same lands during the monsoon season by growing short stiff-strawed nitrogen-responsive varieties in combination with nitrogen fertilizer and good cultural and management practices. These results have been repeated in many tropical Asian countries.

In Japan, lands produce a rice crop every year, but rotation with soil-building crops and other forms of fertilizing procedures are used and adequate irrigation water is provided. Under these conditions, yields hold up very well. In Spain, Italy, and Portugal, rice is grown in rotation with pasture and green manure crops combined with heavy applications of fertilizer on both the rice and pasture crops.

In the United States, the frequency of rice in the rotation system varies with the soil type and with the economic conditions affecting the price or demand for rice. Rotation systems include a wide variety of other crops. As a general rule, heavy clay soils are cropped more frequently than are lighter soils. Since acreage controls were enacted in 1955, fewer rice crops are grown in the rotation systems throughout the country. More emphasis is being placed upon those practices which result in higher acre yields of rice and more efficient production.

Since the beginning of rice production in Texas and Louisiana, it has been a common practice to grow 1 or possibly 2 rice crops, followed by one or more years of grazing beef cattle on the vegetation volunteering between the rice crops. With this system little effort was made to drain the fields between the crops. As better drainage facilities became available, more acres were put into improved pasture in the off-years. While unimproved pasture rotation systems are still more common, improved pastures are gradually increasing in use. In Texas and Louisiana, rice following improved pastures yielded 648 to 810 lb more rice than crops following unimproved pastures, using similar applications of fertilizer (Evatt and Weihing 1957; Walker and Miears 1957).

Improved pastures are usually followed by two or more rice crops. Due to the high cost of developing improved pastures, fields are cropped to improved pasture at least three years before planting rice. Costs of converting rice lands to improved pastures are reduced by seeding pasture crops immediately following rice without seedbed preparation (Moncrief and Weihing 1950). The rice irrigation and drainage system is maintained when the land is in improved pasture. The leading crops for improved pastures, according to Weihing *et al.* (1950), include Bermudagrass, Dallisgrass, and White, Persian, and Hop clovers. Hay crops such as lespedeza, Alyce clover, and rye grass are sometimes planted in rotation with rice, particularly on the better-drained soils. Soybeans are occasionally grown in rotation with rice in Texas and Louisiana.

The leading crops grown in rotation with rice in Arkansas are soybeans, lespedeza, and oats (Jones *et al.* 1952; Mullins and Slusher 1951). They are usually grown as cash crops, but soybeans and lespedeza are sometimes used as green manure crops. Rice is grown on alternate years or every third year, depending upon economic conditions and the fertility level of the soil. Unimproved pasture and summer fallow rotations are occasionally used in Arkansas.

In recent years soybeans have become 1 of the 3 major cultivated cash crops in Arkansas. Much of the same equipment is used for both rice and soybean production. As a result soybeans and rice comprise the main rotation system, with 2 yr of rice followed by 2 yr of soybeans being the most common practice. Rice in rotation with fish production is practiced on areas used as reservoirs in Arkansas. The land is kept flooded for two seasons and stocked with fish. Two rice crops then follow (Green and White 1963).

As late as 1950, no definite rotation system was used in California (Jones *et al.* 1950). It is difficult to grow cultivated crops following rice in California because the soils are heavy clays that are difficult to work following the long flooding period. Wheat and barley, the

FIG. 33.   LEVELING LAND IN PREPARATION FOR RICE PLANTING

crops grown on rice lands prior to rice culture in California, are sometimes grown in a 3-yr rotation with rice, with 1 yr of summer fallow following the rice crop.   Rice and summer fallow in alternate years are also used.   In such cases, the land is spring-plowed following rice and then summer-fallowed until prepared for rice the following spring.   In some cases, the land is not summer-fallowed and the volunteer vegetation is grazed by cattle or sheep.

Field Irrigation and Drainage Systems

One of the first steps in preparing land for rice production in the United States is the laying out of field laterals and drainage ditches. It is essential that irrigation laterals and drainage ditches be situated properly if they are to afford uniform submergence and timely drainage of fields.   Contour levees which divide the fields into subfields or paddies are located by survey and usually placed at grade differences of $2/10$ or $3/10$ of a foot.   Contour levees are located after the land has been plowed and leveled.

The contour levees are constructed in several ways, depending upon the soil type and the irrigation practices to be used.   On the lighter soils in Arkansas, Louisiana, Mississippi, and Texas, disk levee builders are used.   They require two or more passes to form a suitable levee.   On the heavier soils in Louisiana and Texas, hydraulically controlled steel blade pushers are used.   These pushers are custom built in local machine shops.   The levee area is

Fig. 34. THE DISK TYPE LEVEE BUILDER IS WIDELY USED FOR CONSTRUCTING CONTOUR FIELD LEVEES ON THE LIGHTER SOILS FOUND IN THE SOUTHERN STATES

Fig. 35. HYDRAULICALLY CONTROLLED STEEL BLADE LEVEE PUSHERS ARE USED FOR CONSTRUCTING CONTOUR FIELD LEVEES ON THE HEAVIER SOILS

plowed to the center using a moldboard or disk plow before pushing. A second plowing and pushing is required in most cases. When possible, levees should be formed in the fall or early winter to prevent washouts when the fields are flooded.

Extremely large pushers or dikers are commonly used in California. These form a levee in 1 or 2 operations. They require much power or the equivalent of two large track-type tractors. In California, the levees are usually 14 to 16 ft wide, including bar-ditches, and 3 to 4 ft high. In the Southern States, levees are seldom over 10 to 12 ft wide and 2 ft high.

Land Preparation

Rice lands may be plowed in the summer, fall, winter, or even in the spring just prior to seeding. Summer and fall plowing are practiced when land cannot be used profitably for growing another crop. In the southern states, some of the heavier soils are plowed when wet, particularly during seasons of above normal rainfall.

Much rice land is leveled using land planes up to 60 ft in length. Where there is considerable slope to the land, shorter planes are used. Land leveling has proved to be extremely profitable, since level land is easier to drain and irrigate. Rice fields should be drained at the time the young rice seedlings are emerging from the

Fig. 36. Heavy Off-set Disks Are Frequently Used in Preparing Land for Rice Production

soil and again when the fields are ready for harvest.   Poor stand establishment results from improperly drained fields.   It is important to have a uniform depth of irrigation water on the fields to control grass and weeds.

Plowed lands are made ready for seeding by disking and harrowing. Heavy offset disks perform best on the heavier type of soil.   Custom built spiketooth harrows made from pipe are frequently used along with springtooth harrows as a final operation prior to seeding. Heavy clay soils are usually left rough or slightly cloddy, since finely prepared seedbeds are more likely to crust following flooding or rains.   Crusting tends to stimulate the germination of small seeded grasses and other weeds.

Seeding Procedure

Most of the US rice crop is sown in April and May.   In Texas and Louisiana, where a longer growing season prevails, some rice is sown in early March and some after mid-June.

A wide range of seeding methods is used.   In California, the airplane is used for seeding practically all of the crop.   This has been the practice for over 30 yr.   In the southern states, conventional grain drills and the airplane are both used.   The airplane is gaining popularity in the southern states but is probably used for seeding less than 1/2 the acreage.

Rice is sown in both flooded and dry seedbeds.   When sown in flooded fields, the airplane is used exclusively; but, for sowing on dry land, either the grain drill or the airplane can be used.   Dryland seedbed preparation and seeding operations follow closely those used with other cereal crops.   When the airplane is used for dryland seeding, the seeds are usually covered by means of a light spiketooth harrow.

Water-seeding methods vary greatly in the different states.   The procedure was first used in California as a means of controlling water grasses (*Echinochloa* and *Leptochloa* species).   The California method consists of holding flood water from 6 to 8 in. deep on the fields until the rice seedlings have emerged through the water (Finfrock and Miller 1958).   The water depth used depends upon the kind and abundance of water grasses.   It is important to maintain adequate and uniform water depths for the first 21 to 28 days following seeding, as the weeds can emerge through shallow depths of water.   Under California conditions, the deep water also serves as a temperature regulator and thus compensates for the wide variation between prevailing day and night temperatures.   After the rice seedlings emerge, an average water depth of 5 to 7 in. is usually maintained throughout the growing period. In cases of excessive vegetative growth by the rice, which may occur on highly fertile land, the fields may be drained when the rice is 60 to 75 days old (Davis 1950). Herbicides such as Propanil are widely used to control water grasses.

The seeding procedures followed in the southern states depend upon the type of soil, the weed and grass problem, and seasonal conditions. Where weeds and grass are not serious, and weather conditions permit, rice is often sown with a grain drill in rows 7 to 8 in. apart. In clay soils, shallow seeding depths are used and the fields are irrigated following seeding. In sandy loam, the rice is sown two or more inches deep in moist soil and the seedlings emerge without irrigation.

Broadcast seeding on dry soil by airplane or various types of broadcast seeders is frequently practiced on the heavier soils. When sown in this way, seedbeds are left rough and harrowed with a light-weight spiketooth following seeding but prior to irrigation. Irrigation water is applied and drained off as soon as possible. Thorough

Courtesy of IRRI

FIG. 37. THE COMB HARROW IS THE FINAL OPERATION IN LAND PREPARATION
IN THE PHILIPPINES

drainage of fields is essential, as rice seeds covered with soil cannot emerge through water.

Experiments by the IRRI over several years failed to demonstrate any advantage, in terms of grain yield, of transplanting compared to drilled or broadcast planting. Weed control is more difficult with direct-seeded crops because early flooding to smother the weeds is not possible and the young seedlings do not compete as effectively with the weeds as do the larger transplanted seedlings. Transplanting also has the advantage that, under continuous cropping conditions, the fields are available for maturing crops while new seedlings are being raised in the seed bed.

Water-seeding is used in the southern states for controlling certain water grasses and red rice, a variety which volunteers from one rice crop to the next by remaining viable in the soil. In the southern states, rice has more difficulty emerging through water than it does in California due to the warmer water, cloudy weather, seedling diseases, varieties used, and possibly other factors. In Louisiana and Texas, the flood water is drained off immediately following seeding, or it is dropped to shallow depths, until seedlings have emerged and the rice root systems are firmly established in the soil.

The method of water-seeding commonly used in Texas consists of dropping dry or sprouted seeds by airplane into flooded fields that have been harrowed or slightly puddled prior to seeding (Anon. 1951). Light-weight spiketooth or springtooth harrows, pulled by wheeled tractors operating in shallow flooded fields, are used for this operation. Seeds are dropped 12 to 36 hr after puddling, and fields are drained 12 to 36 hr after seeding. The young seedlings develop rapidly under such conditions, ahead of the weeds and grass. The fields are again flushed a week to 10 days later, depending on the weather conditions and the growth of the rice. The fields are flooded as soon as weed and grass seedlings emerge, which may be before the rice seedlings reach a height of 3 to 4 in. If weeds are not a serious problem, it is better to wait until the rice seedlings are 5 to 6 in. tall before holding a flood, regardless of the method of seeding used.

In some sections of Arkansas, Louisiana, and Texas, water-seeding is used and the land is not worked following irrigation. The water is held for several days after dropping sprouted or dry seed. In most cases, the fields are then drained or the water is lowered to shallow depths to enable the root systems of the young seedlings that are on the surface of the soil to become established.

In the southern states, wet seedbed preparation or water cultivation is sometimes necessary during seasons of heavy rainfall. Disks

Courtesy of IRRI

FIG. 38. PLOWING WITH A SMALL TRACTOR IN THE PHILIPPINES

and harrows are used to work the land under flooded conditions, and seeds are then dropped by airplane in flooded fields.

Sprouted seeds, when used in water-seeding, should have the sprout just breaking the husk. If the sprouts are longer, they are likely to be damaged in handling or in seeding. Seeds are sprouted in small water-tight bins or in sacks placed in canals or ditches for 12 to 48 hr followed by a draining period of 12 to 24 hr. The soaking and draining period is dependent upon prevailing temperatures. It may be necessary to sprinkle or pour water over the rice to lower the temperature during warm weather. In California, soaking vats are generally used for sprouting seeds (Davis 1950).

Much of the seed rice used in the southern states is treated with fungicides. The most commonly used chemicals are Ceresan, Arasan, and liquid mercurial seed treatments. Insecticides such as DDT, aldrin, and dieldrin are sometimes used as seed treatments in the southern states to control insects attacking the rice crop in the seedling stage (Bowling 1957).

Seeding rates vary from 80 to 140 lb per acre in the southern states (Jones *et al.* 1952). Seeding rates as low as 60 lb per acre have been used in the western part of the Texas rice area (Hodges 1957). Seeding rates vary from 125 to 150 lb per acre in California.

Fertilizers

Commercial fertilizers are widely used throughout the rice regions of the United States. The use of commercial fertilizers dates back to the early 1900's, when basic slag and superphosphate were used to a limited extent on relatively new lands in the southern states. Prior to World War II, the importance of nitrogen fertilizers was recognized (Wyche 1941; Davis and Jones 1940) and today large quantities of nitrogen and phosphate fertilizers, as well as a limited amount of potash, are used.

In Texas and Arkansas up to 80 lb of nitrogen per acre are widely used along with approximately 40 lb of phosphate. In Louisiana and Mississippi, slightly lower nitrogen rates are employed. When very early maturing varieties (about 100 days from seeding to maturity) are being grown, as much as 120 lb of nitrogen may be used. In California, 60 lb of nitrogen per acre are generally used on soils of average fertility. Lower amounts of fertilizer are required on the more fertile soils. Rice responds best to ammonia forms of nitrogen, a fact that is recognized throughout the world.

In the southern states, most fertilizers are broadcast by airplane after the rice has been sown, as better weed and grass control is obtained by fertilizing after the rice has emerged from the soil and just ahead of the first flood water. Reynolds (1954) recommends that all fertilizer be applied as a top-dressing 35 to 40 days after seeding where grass and weeds are troublesome. In California, most of the fertilizer is broadcast just prior to seeding.

Weeds

Weeds must be constantly guarded against in US rice fields. Control practices such as rotation systems, grazing, summer tilling, water-seeding, and timely irrigation and drainage have long been used in the rice-growing regions. Until the development of phenoxy herbicides after World War II, broad-leaved weeds were difficult to control on heavily cropped lands, particularly in the southern states. Many farms were so badly infested that it was no longer profitable to grow rice. Today, many of the water-loving broad-leaved weeds so common in rice fields a few years back are no longer a threat to the rice industry. The most common rice weeds are shown in Table 50.

The phenoxy herbicides are relatively ineffective against grass type weeds. In about 1962 the widespread use of Propanil began and by 1965 as much as 90% of the US rice crop was sprayed with Propanil (Smith 1966). New chemicals which are proving effective

TABLE 50

THE MORE COMMON WEED PESTS FOUND IN UNITED STATES RICE FIELDS[1]

Common Name	Scientific Name
Grasses	
Red rice	*Oryza sativa*
Big barnyard (Cockspur)	*Echinochloa crusgalli*
Little barnyard (Jungle rice)	*Echinochloa colunum*
Spangletop	*Leptochloa fasicularis*
Rice cutgrass	*Leersia oryzoides*
	Leersia hexandra
Large crabgrass	*Digitaria sanguinalis*
Jointgrass	*Paspalum distichum*
Longtom	*Paspalum lividum*
Sedges	
Bulrush	*Scirpus fluviatilis*
Elegant cyperus	*Cyperus sabulosus*
Yellow cyperus	*Cyperus iria*
Yellow nutgrass	*Cyperus esculentus*
Jointed sedge	*Cyperus articulatus*
Umbrella plant	*Cyperus verins*
Spikerushes	*Eleocharis* spp
Spearhead	*Rhynchospora corniculata*
Hurrah grass	*Fimbristylis miliaceae*
	Fimbristylis puberula
Leguminous weeds	
Tall indigo	*Sesbania macrocarpa*
Coffee bean	*Sesbania drummondi*
Spunk bean	*Sesbania vesicaria*
Curly indigo	*Aeschynomene virginica*
Sickle senna	*Cassia torra*
Other weeds	
Alligator weed	*Alternanthera philoxeroides*
Arrowhead	*Sagittaria* spp
Batwing (Day flower)	*Commelina communis*
Cattail	*Typha latifolia*
Curly dock	*Rumex crispus*
Eryngo (Star thistle)	*Eryngium hookeri*
Large buttonweed	*Diodia virginiana*
Mexican weed (Birdeye)	*Caperonia palustris*
Mud plantain	*Heteranthera limosa*
Redstem	*Ammania coccinea*
Redweed (Goat's beard)	*Melochia corchorifolia*
Seacoast sumpweed	*Iva ciliata*
Seaweed	*Sphenoclea zeylanica*
Sida or ironweed	*Sida* spp
Slim aster	*Aster exilus*
Snow-on-the-prairie	*Euphorbia bicolor*
Water hyssop	*Bacopa rotundifolia*
Water plantain	*Alisma plantago*
Water primose	*Jussiaea* spp
Wooly croton (Goatweed & Turtle back)	*Croton capitatis*

[1] Compiled from information reported by Davis (1950), Hodges (1957), and Williams (1955 and 1956), and from a weed collection maintained at the Rice-Pasture Experiment Station, Beaumont, Texas, which was collected by Mrs. Betty Higginbotham (unpublished).

against grasses and other weeds are Shep, Molinate, and herbicide 0-44 (Smith 1968).

Both ground and airplane spraying of herbicides are practiced, but the airplane is much preferred where it can be used without damaging nearby sensitive crops.

Red rice is a serious weed pest in many sections of the rice growing states.   It resembles cultivated varieties in many respects but tillers more profusely and shatters easily.   The grains show a marked dormancy following ripening, and as a result, the grains of red rice can remain viable in the soil for several years.

Diseases

There are several diseases affecting the rice plant in the southern states, but they seldom cause heavy losses.   The only serious disease in California is seedling blight.

One of the most serious diseases in the southern states is straighthead, a physiological disorder (Atkins 1958; Tisdale and Jenkins 1921) which is most prevalent on lighter soils or where large quantities of organic matter have accumulated in the soil.   Straighthead is characterized by panicles that remain erect because the grains have failed to develop properly.   The panicles may be distorted and glumes deformed or, in extreme cases, the plants may fail to head. The disease can be controlled or greatly reduced by draining fields at about the jointing stage of growth (Cheaney 1955) and by growing resistant varieties (Atkins *et al.* 1956).

A number of leaf spot diseases (*Helminthosporium oryzae, Cercospora oryzae, Entyloma oryzae*) caused by fungi are common in the southern states, but they seldom cause losses of economic importance (Atkins 1958).   Stem rot (*Sclerotium oryzae*) occasionally causes rather severe losses from lodging and from failure of grains to fill properly.

Blast, one of the most destructive diseases of rice, is caused by the fungus *Piricularia oryzae*.   It can attack plants either at the seedling stage (leaf blast) or at time of flowering (neck rot).   Different races predominate in different regions, and they are identified on the basis of varietal susceptibility or resistance.   The varieties of rice now grown in the United States are moderately resistant to blast but on new lands, or when extremely high levels of nitrogen are present, the disease can cause heavy losses.   This is particularly true during periods of cloudy, humid weather.

White tip, a disease caused by a seedborne nematode (*Aphelenchoides besseyi* Christie), causes rather severe losses on susceptible

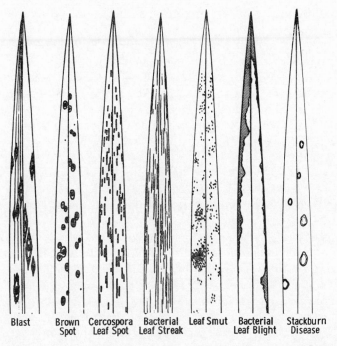

Blast   Brown   Cercospora Bacterial Leaf Smut Bacterial Stackburn
         Spot   Leaf Spot   Leaf Streak   Leaf Blight Disease

Courtesy of IRRI

FIG. 39. TYPICAL LEAF LESIONS OF SEVEN RICE DISEASES

varieties (Atkins 1958). Most of the varieties now grown are resistant or moderately resistant to white tip. The most satisfactory
control measures are hot water treatment of seed (Cralley 1949),
along with the use of resistant varieties.

Some of the virus diseases of tropical Asia are grassy stunt, orange
leaf, yellow dwarf, tungro, and transitory yellowing. Most of these
are known by different names in different countries, and the identification of distinct diseases may be difficult, especially since the
symptoms may also vary from country to country. The viruses are
transmitted by insect vectors (such as the rice green leafhopper,
Nephotettix impicticeps), most of them quite specific. Control measures include use of resistant varieties and control of the vectors.

Hoja blanca, a virus disease occurring in Central and South
America, has also been observed in Florida and Louisiana (Atkins
and Adair 1957), but has not caused damage to the crop in the important rice producing states. It has caused heavy losses in Cuba,
Venezuela, and Colombia. Varieties resistant to hoja blanca are

available, but the leading long grain varieties in the United States are susceptible to this disease. The virus of hoja blanca is carried by a leafhopper. The hoja blanca disease resembles in some respects the stripe virus disease (*Oryza* virus 2) occurring in Japan.

Bacterial leaf diseases which occur in most tropical Asian countries are bacterial leaf blight (*Xanthomonas oryzae*) and bacterial leaf streak (*Xanthomonas translucens* f. sp. *oryzae*). Neither of these diseases has been positively identified as occurring in the Western hemisphere.

Insects

Several species of field insects attack the rice plant, particularly in the southern states. The most common rice insect pests are rice water weevils, army worms, rice stalk borers, rice stinkbugs, leaf miners, sugarcane beetles, and grasshoppers. The rice water weevil, or root maggot (*Lissorhoptrus oryzophilus*), attacks young seedlings by feeding on their leaves in the adult stage and by feeding on the roots in the larval stage. The adults deposit their eggs in the roots of the rice plant immediately after the first flood water has been applied. The field must remain flooded for the larval stage (maggots) to develop. The maggots feed upon the young roots and cause serious damage to the root system. Control measures are: timely draining of fields (Isely and Schwardt 1934), the use of insecticides such as aldrin or dieldrin for treating the seed (Bowling 1957), and spraying fields with insecticide (Whitehead 1954). In less than ten years the water weevil has developed resistance to dieldrin so systemic insecticides are being considered (Bowling 1968).

Army worms cause serious damage to rice by devouring the leaves and stems. They are easily controlled by aerial application of various insecticides. The two stalk borers affecting rice in the United States are the sugarcane borer (*Diatrae saccharalis*) and the rice stalk borer (*Chilo plejadellus*) (Douglas and Ingram 1942). Heavy losses seldom occur, but borers are found in most rice fields. They feed within the rice stems during the jointing and heading stages of development and cause the panicles to dry up and turn white. This usually occurs before the grains are fully developed. In tropical countries with year around growing seasons, stem borers cause heavy losses. Control is by means of systemic insecticides.

The rice stinkbug (*Solubea pugnax*) and several other related species cause serious losses to the developing rice grain by puncturing the kernel in the milk or the dough stage. Losses result from grain failing to develop and from the discoloration caused by the puncture and the accompanying fungal infection on kernels which do develop. Many of these damaged grains are poorly filled and break up in the milling operation, thus reducing head, or whole grain, yields. Some punctured grains develop normally and do not break in milling, but a reduction in quality and grade results from the discoloration. Such kernels are commonly referred to in the rice trade as "pecky rice." Stinkbugs and similar insects can be controlled by airplane spraying of recommended insecticides such as aldrin, dieldrin, and toxaphene (Bowling 1956).

Grasshoppers cause damage to rice by feeding upon the glumes and other floral parts of the developing rice grain. They are also controlled by airplane spraying with insecticides.

Leaf miners affect rice plants in the seedling stage and, in cases of heavy infestation, they cause serious reductions in stands.

The sugarcane beetle, *Euetheola rugiceps*, is most severe in southwest Louisiana but also occurs in the other southern rice-growing states, according to Douglas and Ingram (1942). The adult beetles attack rice plants in both the seedling and the mature plant stages by feeding on the lower stems when the fields are not flooded.

The difficulty in reaching larvae concealed within the plant, and of maintaining adequate concentrations of insecticide on the plant throughout the growth cycle, has directed attention away from foliar sprays in the direction of systemic insecticides such as lindane (the gamma isomer of benzene hexachloride). It has been shown by IRRI research workers that stem borers can be controlled by two applications of 2 to 3 kg per ha of lindane to the paddy water 50 and 80 days after transplanting. This treatment is, however, comparatively ineffective against the pink borer, and does not control the green leafhopper. Use of carbaryl at 1 to 2 kg per ha in conjunction with lindane gave protection against heafhoppers. Diazinon seemed to be almost as effective as the lindane and carbaryl combination in preliminary tests. The question of whether or not the residues would be within acceptable limits for the US trade when the insecticides are used as suggested has not been resolved.

Harvesting

Rice is harvested from late July to late October in Texas and Louisiana. In Arkansas and California the harvest season is shorter due to the shorter growing season.

Self-propelled combines are commonly used for harvesting rice in all of the rice-growing regions. Conventional combines are used throughout the southern states and parts of California, with minor

FIG. 40.   SELF-PROPELLED COMBINE AND SELF-PROPELLED CART

changes to adapt them to muddy harvesting conditions.   In California large custombuilt full-track combines are sometimes used (Davis 1950).   These machines are built in farm shops in the rice areas.   Pickup reels are now standard equipment on most rice combines as lodging is frequent, especially where high rates of nitrogen fertilizers are used.

Rice is usually harvested when the moisture content of the grain is between 18 and 25%.   Therefore, it is necessary to dry the grain artificially before placing it in permanent storage.   If the rice is harvested at higher than 20 to 25% moisture content, there may be immature grains on the lower portions of the panicles which will be chalky and may break in milling.   When harvested at moisture contents below 18%, the grains are more apt to check or fracture, resulting in reduced whole grain yields when milled.   Checking of grains is caused by wide variations in temperature when accompanied by alternate wetting and drying of the rice grains from showers or heavy dews.

As a general rule, rice is harvested at higher moisture contents in California than in the southern states because the shorter, fuller grain varieties grown in California are more apt to check or fracture. Also, the wide extremes in day and night temperatures and humidity may lead to fracturing of grains when moisture contents drop below 20%.

Rice fields often are muddy or otherwise inaccessible to motor trucks at time of harvest. Therefore, some means of conveying the threshed rice from the combine to the trucks must be used. For this operation various types of conveyor bins, tanks, or carts are used. The most popular means of conveyance in recent years has been self-propelled carts or tanks, custom-made in farm shops. In California track-type tanks are sometimes used, but in the southern states wheel-type carts are more general.

DRYING AND STORAGE

Combine harvested rice is cut at a relatively high moisture content and consequently must be dried before it can be placed in permanent storage. Since rice is marketed as a whole grain product, it is important that the grains not be fractured or otherwise damaged before or during the drying process. Large column-type, continuous-flow driers using heated air as well as batch-type units have been widely used.

In recent years, "on the farm bin drying," with unheated air, has been successfully used throughout the rice-growing states (Morrison *et al.* 1954; Sorenson and Davis 1955; Hildreth 1955).

Some rice has also been dried in sacks in tunnel driers, using heated air. Today, due to high handling costs, sack drying is for the most

FIG. 41. RICE-DRYING PLANT SHOWING THE LARGE COLUMNAR DRIERS AND A QUONSET STORAGE BUILDING

part limited to drying seed rice.   It affords a means of drying with a minimum chance of mixing, which is vitally important in maintaining high quality seed rice.

The column-type methods of drying rice consist of passing heated air through falling or stationary columns of rice for short periods of time (Wasserman *et al.* 1966).   Usually two or more passes through the drier are required to bring the moisture content down to 12.0 to 13.5%, which is usually considered a safe range for storage.

The maximum temperature of heated air used in column-type driers seldom exceeds 140°F.

Most column-type units are relatively large commercial plants and frequently dry over 120,000 cwt of rough rice annually.   Considerable storage space is required to handle such quantities of rice, even though a major part of the grain may be shipped from the drier as soon as it has been dried.   In order to make more efficient use of drying facilities, most commercial drying plants now have ventilated holding bins that allow high-moisture rice to be held for longer periods before drying.

Bin drying has been increasing in popularity in recent years. Both round bins and quonset-type bins have been used with equal success.   Large volumes of air must be forced through the rice when it is to be dried in this way.   The minimum airflow rate recommended by Sorenson and Davis (1955) was 9.0 cfm per barrel (162 lb) forced through an 8-ft depth of rice.   The rice is dried to 13% moisture or less.   It is of interest to note that this method of drying can be successfully used in the Gulf Coast areas of Texas and Louisiana where periods of relatively high humidity are not uncommon.   Under these climatic conditions, the air is heated a few degrees to increase the drying rate.

Practically all of the US rice crop is stored in bulk.   Prior to World War II, most of the Texas and Louisiana crop was sacked as it was threshed, using stationary threshing machines.   The grain was stored in sack warehouses.   However, when combine harvesting and artificial drying replaced the binder-thresher method of harvest, there was a rapid shift to bulk storage.   Varied and different types of structures are used including wood, concrete, and steel bins.

The maximum moisture content for safe storage of rough rice is approximately 12% in any part of the bin, according to Sorenson and Davis (1955).   Moisture contents as high as 14% for the wettest grain in the bin have been recommended for temporary winter storage (Anon. 1950).   Forced ventilation is widely used throughout the rice-growing regions for keeping stored rice in condition in both on-the-farm and commercial storage units.

Courtesy of IRRI

FIG. 42. THRESHING SCENE IN THE PHILIPPINES

VARIETIES

The varieties of rice grown in the United States are comparatively short-strawed, intermediate tillering types. This is in contrast to the lush-growing, profuse-tillering, long season varieties grown in the tropics. United States varieties can be divided into early, mid-season, and late maturing groups based on the number of days from seeding to maturity. Early maturing varieties mature in 100 to 120 days, midseason maturing varieties in about 140 days, and late maturing varieties in about 170 days. Today, most of the acreage is sown with early and midseason maturing varieties. It is only in Texas and Louisiana that the late maturing long grain varieties are grown. Even there they are of only minor importance because of the long watering period and because they seldom yield as high as the early and midseason maturing varieties. Late maturing varieties, such as Rexoro, Texas Patna, and TP 49, are better suited for par-boiling and canning than other varieties because the grains remain firmer and show less splitting when processed. This is thought to be an heritable character and has been transferred to early maturing long grain varieties such as Belle Patna, Bluebelle, and Dawn. These earlier maturing varieties have largely replaced the late maturing varieties.

The varieties of rice grown in the southern states, with but few exceptions, differ from those grown in California. This is due to the fact that the two regions have vastly different climatic conditions and require varieties of different adaptation.

In California, the short-grain *japonica*-type varieties are grown because they are hardy types which have the ability to emerge through several inches of relatively cold water when sown in flooded fields. They are also able to withstand the extremes in temperatures prevailing in the interior valleys of California. Under California conditions, they are well adapted to mechanized harvesting methods.

The varieties of rice grown in the United States have, for the most part, been developed from *japonica* and *indica* varieties originally introduced from Japan, the Philippines, and Formosa. From these original introductions were selected such varieties as Blue Rose, Early Prolific, Caloro, Colusa, Fortuna, Nira, and Rexoro (Jones *et al*. 1953).

The four US rice experiment stations located at Stuttgart, Arkansas; Biggs, California; Crowley, Louisana; and Beaumont, Texas have had active breeding programs for many years. They are operated cooperatively by the four state agricultural experiment stations, the US Dept. of Agr., and local industry organizations. All of the US rice crop, with the exception of a negligible acreage, is now sown to varieties developed by these organizations.

Hybridization programs conducted at the US rice experiment stations have brought about the development of new varieties by intercrossing commercially grown varieties and foreign introductions. Varieties that have been so developed include Bluebonnet, Bluebonnet 50, Sunbonnet, Toro, Belle Patna, Bluebelle, Dawn, Starbonnet, Century Patna 231, Texas Patna, TP 49, Nato, Magnolia, Saturn, Northrose, and Calrose. These varieties now constitute approximately 90% of the US rice crop. The only major varieties not the result of hybridization are Caloro and Colusa, leading California varieties, and Rexoro, a late-maturing long slender grain rice grown in Texas and Louisiana.

Some of the varieties of rice grown in the United States were developed by crossing *japonica* and *indica* varieties in order to combine the desirable features of both groups. Varieties developed from such crosses are Century Patna 231, Toro, Belle Patna, Bluebelle, Starbonnet, Saturn, Nato, and Magnolia. The *indica* varieties used were, for the most part, long-grain, and the *japonica* varieties were short-grain types.

The *japonica* varieties are usually recognized as being higher yielding than the *indica* types (Ramiah and Rao 1953; Ghose *et al.* 1956). The *japonica* varieties have small stems, relatively short straw, and narrow leaves of short to intermediate length. They usually respond well to heavy applications of nitrogenous fertilizers. The straw of the *japonica* varieties is tough and willowy and not so desirable for combine harvesting as the more frangible and stiffer straw of certain *indica* varieties. Also, the *japonica* varieties do not possess the range of grain size and shape or the processing and cooking characteristics needed in rice for the US market.

Most of the *japonica*-type rices are sensitive to length of day under US climatic conditions, while certain *indica* types have a more or less fixed growth period (Beachell 1943). By using varieties of a fixed maturity period, the harvest season can be spread over a longer time, resulting in more efficient use of expensive harvesting and drying facilities.

Most of the long grain varieties now grown in the southern states show a partial dormancy of the embryo at the time the grain ripens, and consequently the grains seldom germinate on the panicle during periods of rainy or humid weather. *Japonica* varieties seldom show dormancy. Dormancy, when present, lasts for a month or more after harvest, and it is not uncommon for lots of long grain varieties having high germinating potential to germinate as low as 50% if tested shortly after drying.

The *japonica* varieties are, for all practical purposes, no longer grown in the southern states because they are not adapted to combine harvesting methods under prevailing conditions. They produce high yields but lodge severely and are difficult to thresh. The inclement weather prevailing throughout the southern states during the harvest season requires that a variety not lodge excessively and that the grain thresh freely from the panicles. In California, where the harvest can be completed with little or no rainfall, lodging is not so critical. However, it should be pointed out that, under California conditions, varieties such as Caloro and Calrose grow much shorter than they do in the southern states and are less likely to lodge.

The earlier maturing long- and medium-grain varieties grown in the southern states are not adapted to the extreme temperatures and low humidity of California. While they will mature grain under the West Coast conditions, experimental yield trials reported by Jones *et al.* (1952) show the yields to be relatively low compared with the adapted *japonica* varieties.

The continual world wide shortage of rice may be somewhat alleviated by new high-yielding varieties which have recently come into existence as the result of the work of plant breeders. A high yielding *indica* strain designated IR 8, developed at the International Rice Research Institute in the Philippines, has been described as giving higher yields than *japonica* varieties in the tropics. Seed became available in moderate quantities in 1966, and practical yields are apparently bearing out the predictions of very high yields that were made on the basis of results in test plots. The new strain is said to have strong seedling vigor, high tillering ability, short stature (90 to 105 cm), lodging resistance, good response in terms of grain yield to high levels of nitrogen, moderately early maturity (about 120 days from seedling to harvest), insensitivity to photoperiod, moderate seed dormancy, and moderate resistance to tungro virus. The plants are susceptible to bacterial leaf blight and to some races of rice blast fungus, but are apparently no worse in these respects than some common varieties. The grain is medium long and bold, chalky, and prone to breakage. The head rice recovery from milling is therefore relatively low. IR 8 is high in amylose and has a low gelatinization temperature. Acceptability of the cooked rice is adequate for consumers in most parts of tropical Asia. Other short, stiff-strawed nitrogen-responsive varieties of improved disease re-

Courtesy of IRRI

FIG. 43. A PLOT OF IR8. NOTE SHORT STURDY STEMS, ERECT GROWING LEAVES, AND HIGH TILLERING

sistance and having better milling and cooking characteristics are being developed at IRRI and by government sponsored breeding programs in several other tropical Asian countries.

STRUCTURE AND COMPOSITION

The composition of typical samples of rice and rice products is shown in Table 51. The changes in composition as the grain is milled reflect the uneven distribution of the components relative to the distance from the surface. As the layers are successively removed, the proportion of proteins, fats, and vitamins in the remaining kernel decreases, while the proportion of carbohydrate increases.

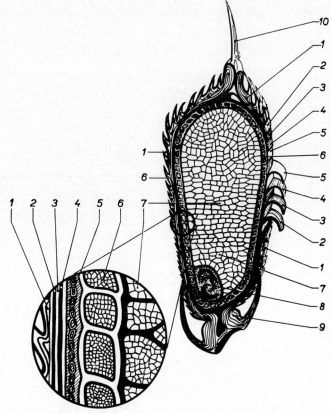

Courtesy of FAO

FIG. 44. STRUCTURE OF THE RICE KERNEL

(1) Hull (glume and palea), (2) epicarp, (3) mesocarp, (4) cross layer, (5) testa, (6) aleurone layer, (7) starchy endosperm, (8) embryo, (9) nonflowering glumes, and (10) apex or beard.

TABLE 51

COMPOSITION OF RICE AND RICE PRODUCTS[1]

	Brown Rice	Rice Bran	Rice Polish	Milled Rice	Parboiled Rice
Carbohydrates, %[2]	87.2	46.6	66.8	91.5	. . .
Protein, %[2]	8.3	14.6	13.2	7.6	. . .
Fat, %[2]	2.0	13.4	10.7	0.3	. . .
B-Vitamin, μg/gm					
Thiamine	4.2	27.9	23.9	0.80	2.57
Niacin	47.2	408.6	384.7	18.1	39.8
Pyridoxine	10.3	32.1	30.8	4.5	. . .
Pantothenic acid	17.0	71.3	92.5	6.4	. . .
Riboflavin	0.53	2.22[3]	1.34	0.26	0.36

[1] See table compiled by Kester (1959) for the original references. Carbohydrates for bran and polish are the percentages of nitrogen-free extract.
[2] On moisture-free basis.
[3] Average of first- and second-break brans.

Starch

Milled rice will contain from about 84 to over 90% starch (dry basis). This component is in the form of angular granules measuring about 2 to 7 μ in their largest dimension. Many of these particles are roughly pentagonal in outline. In size and shape, they are like the granules of oat starch, but differ in that few if any rounded particles occur. The hilum is centric and indistinct, and birefringence is weak.

In the seed, starch exists in compound granules containing up to about 150 of the smaller particles. The clusters are firmly bound together by a water-insoluble protein matrix. The protein can be dispersed and the granules released by various alkaline steeping methods.

Samples of commercial rice starch obtained from US sources lose their birefringence between 144° and 169°F and gelatinize in the range of 154° to 172°F when cooked in water on a Kofler hot stage (Schoch 1967). The viscosity behavior is similar to that of corn starch. Nonwaxy starch from *indica* rice contains 21 to 33% amylose, while that from *japonica* rice will contain a lower amount (17 to 19%) of the linear fraction. Schoch also states that there is a wide variation in iodine affinity and viscometric molecular weight depending on the variety from which the amylose is taken. For example, Century Patna amylose has an iodine affinity of 18.1% as compared to Caloro's 15.0%; Rexoro linear fraction starch has an apparent molecular weight of 325,000 while the equivalent figure for Century Patna is 100,000, etc.

Rice starch preparations contain small amounts of lipid. Fats determined by acid hydrolysis amount to 0.83% according to Taylor and Nelson (1920) or 0.67% according to Lindemann (1951). A large part of this material consists of fatty acids which cannot be removed by petroleum ether but dissolve readily in methanol or 80% dioxane.

Waxy (glutinous) rice has been grown in Asia for many centuries. The major portion of the crop in Laos and Cambodia, and in parts of Thailand and Vietnam consists of waxy varieties. The principal difference between this type and normal rice lies in the starch, that from waxy rice consisting of almost 100% amylopectin and thus staining red with iodine solutions while starch from other rice contains a substantial proportion of amylose and stains blue. The flour and the purified starch from waxy rice has been found to have advantages for certain food and industrial uses. Gravies, sauces, and puddings based on this starch are very resistant to synersis during freezing and thawing. Deatherage *et al.* (1955) found 2% amylose in the starch of a waxy rice sample from China, while an Asahi Mochi variety grown in this country contained no identifiable amylose.

Most of the protein in rice kernels is present as discrete protein-rich bodies or aleurins in the endosperm, similar to the distribution in many other seeds. This protein is largely insoluble in water, and most of it is of the type called glutelin, that is, it is soluble only in dilute acids or alkalies. The second most abundant type is the salt-soluble globulins. These groups are composed of an unknown number of molecular species. They are generally isoelectric around pH 7 (Williams 1961).

The protein content and the amino acid composition of the protein will vary with growing conditions and variety. Williams (1961) stated that the same variety of rice grown at different locations varied considerably in amino acid composition. Different rice varieties grown at the same location will also vary in their content of amino acids.

The distribution of amino acids is not uniform within the grain. Milling not only causes a decrease in the crude protein of the grain, but also a change in the amino acid pattern of the remaining protein. Generally, the decrease of lysine on milling is disproportionately large, while there is a relative increase of isoleucine, leucine, and valine. Furthermore, as the percentage of nitrogen increases in the grain, due to growing conditions, for example, the tendency is for the percentage of lysine in the protein to decrease.

Table 52 gives recently obtained values for the amino acid composition of one variety of rice. These figures are considered to be fairly representative of brown rice.

TABLE 52

AMINO ACID COMPOSITION OF PROTEIN FROM A BROWN RICE[1]

	Gm/100 Gm Protein		Gm/100 Gm Protein
Alanine	6.46	Methionine	1.91
Arginine	7.18	Phenylalanine	5.60
Aspartic acid	10.2	Proline	4.97
Cystine	1.17	Serine	5.27
Glutamic acid	22.5	Threonine	3.72
Glycine	5.09	Tryptophan	1.35
Histidine	2.24	Tyrosine	3.78
Isoleucine	4.62	Valine	6.41
Leucine	9.08	Ammonia	2.59
Lysine	3.22		

[1] Palmiano *et al.* (1968). Total of the listed values is greater than 100 grams due to analytical uncertainties.

Lipids

Lugay and Juliano (1964) found about 0.36% lipid in milled rice as compared to 21.3% in bran (by weight basis). The glycerides of rice bran oil contain the fatty acids having even numbers of carbon atoms from 14 to 20, as well as chains of 11, 13, and 15 carbon atoms (Sreenivasan 1968). The latter were present in the small amounts of 0.2, 0.6, and 0.9%, respectively. Gas-liquid chromatography indicated that the bran lipids had significantly higher mean contents of linoleic and linolenic acids, but lower contents of myristic, palmitic, palmitoleic, and stearic acids, than the lipids of milled rice. The iodine values of bran lipids were reported to be significantly higher than for milled rice lipids, and the lipids of *indica* varieties had lower iodine values than did this component in *japonica* rice.

Other Constituents

Vitamins (particularly those of the B-complex), sugars, hemicellulose, fiber, short-chain fatty acids, inorganic cations, compounds of phosphorus (particularly orthophosphates and esters of inositol), nucleic acids, pigments, and other substances have been found in rice. Phytin, the principal phosphorus compound in rice, is said to constitute more than 8% of the bran. Calcium, iron, potassium, magnesium, sulfur, phosphorus, and minor amounts of other elements have been found in rice ash.

QUALITY FACTORS

The quality of rice for milling is based on the yield of total milled rice, yield of whole grains, appearance, and suitability of the product for its intended purpose. The rice miller is interested in the variety, of course, but he must also evaluate the incoming grain on the basis of moisture content and freedom from red rice, peckiness, foreign seeds, chalky kernels, soil, insects, and mustiness.   Samples of the rice may be subjected to a pilot milling test to determine the expected yield.   In the United States, practically all of the domestic rice crop is milled to a high degree.   The yield of whole grains, which bring the highest price per pound of the milling products, is affected not only by the milling efficiency and the extent of milling but also by the characteristics of the kernels.

Varietal, environmental, and storage factors affect the mill yields of rice.   Beachell and Halick (1957) reported total milled rice yields varied from 72.6% for Caloro to 65.5% for Century Patna 231, using laboratory test metods.   Whole grain yields (head rice) varied from 60.7% for Toro to 39.5% for Nira.   Some of these differences can be attributed directly to inherent varietal characteristics.

The quality of the milled grain must be related to consumer acceptance, but rice is used for many purposes, and the quality factors looked for in the milled grain will vary according to its intended use. Some of the most important processing methods or applications of rice include: (1) consumer use as boiled rice prepared in the home, institution, or restaurant.   This is by far the predominant consumption route throughout the world;   (2) canned rice—as in soups, rice pudding, Spanish rice, etc.;   (3) brewers' use as a carbohydrate source or "adjunct" in beer;   (4) dry cereals, such as puffed rice;   (5) cereals for home preparation, such as rice granules, drum-dried baby cereals, etc;   (6) industrial uses—preparation of rice starch and few other relatively minor processes;   (7) use of rice flour as a thickener in sauces, gravies, puddings, etc.;   and (8) preparation of certain fermented and fungus-treated foods.

In the following paragraphs, the quality factors affecting the most important of the preceding modes of consumption will be discussed in some detail.   Generally, consideration will be restricted to practices in the United States.

Consumer preferences for grain to be used as steamed or boiled rice vary in different countries.   Waxy or glutinous rice is preferred in Laos and northern and northeastern Thailand.   Filipinos and Indonesians generally prefer rice that remains soft or moist on standing after cooking, and this property has been correlated with the presence of between 17 and 22% amylose in the starch.   In Malaysia, Ceylon, Pakistan, North Vietnam, and parts of Thailand, the preference is for a cooked product with large grain size and a high degree of flakiness.   There are regional differences within the United

States, but most consumers desire a grain that retains its contours and has relatively firm texture when cooked and for as long as possible while being kept warm for serving. Stickiness, slimy free fluid between the grains, and mushy or elastic kernels are definitely negative quality factors.

The cooking behavior of rice is, in large part, a function of the characteristics of the starch, especially as related to amylose content, gelatinization temperature, and pasting response. The protein and lipid components undoubtedly play a role, but the extent of their influence and the mechanism by which they affect the texture is largely unknown. It appears that the cooking time increases with higher protein content while the water absorption decreases. High-protein grain also has a creamier appearance.

Physical and chemical differences in the milled rice grain can be used to some extent in rating the consumer acceptability of rice varieties. A number of the more recently reported tests used for classifying varieties using physical and chemical methods will be discussed below. These differences along with agronomic differences are shown in Table 53.

TABLE 53

COMPARATIVE PHYSICOCHEMICAL CHARACTERISTICS OF SOME RICE VARIETIES[1]

Variety	Amylose Content	Iodine Blue Value	Alkali Reaction	Water Uptake at 171°F	Amylograph-Viscosity	
					Peak	After 10 Min at 201°F
	%	% Transm.	Avg No.	No.	B.U.[2]	B.U.[2]
Long-grain varieties						
Dawn	25.5	14	3.8	121	767	415
Bluebelle	24.1	15	3.7	135	830	433
Belle Patna	23.8	14	3.2	137	828	410
Bluebonnet 50	23.4	16	3.8	153	830	410
Medium-grain varieties						
Saturn	15.0	54	6.2	322	995	417
Nova 66	16.1	50	6.3	351	975	392
Nato	15.6	59	6.3	357	937	405

[1] Based on Webb (1966). See this paper for original references. Milled rice, 11.5% moisture content.
[2] Brabender units.

Differences in gelatinization and pasting characteristics were reported by Halick and Kelly (1959) using a standard model "amylograph." A slurry of 50 gm of ground milled rice and 450 ml water was heated at the rate of 2.7°F per minute. After reaching 201°F, temperature was maintained constant for 20 min, then lowered to 122°F at the rate of 2.7°F drop per minute. The salient varietal differences reported were (1) the temperature at which gelatinization occurred, (2) the peak viscosity, (3) the viscosity at 201°F and, after 20 min at this temperature, and (4) viscosity changes while cooling to 122°F.

The short- and medium-grain varieties such as Caloro, Zenith, and Nato and the long-grain variety, Toro, gelatinized at the lowest temperatures. Century Patna 231 and Early Prolific gelatinized at the highest temperatures and Bluebonnet 50, Sunbonnet, Texas Patna, and Rexoro gelatinized at intermediate temperatures. Gelatinization temperatures were independent of grain type and amylose content, but it was concluded that there was a direct relationship between maximum viscosity of the hot paste and the viscosities when cooling to 122°F. Gelatinization temperatures varied from 149°F for Zenith to 175°F for Century Patna 231. A glutinous variety (no name reported) gelatinized at a temperature of 136°F or lower than any of the common varieties.

Halick and Kelly (1959) also report a rapid method of classifying rice varieties on their ability to absorb water when cooked at different temperatures in excess water. Two-gram samples of whole grain milled rice were cooked at temperatures of 162°, 171°, and 180°F for 45 min in excess water. Results were reported on the basis of the percentage of water absorbed by weight during cooking. The results obtained show this test to be valuable in determining the approximate gelatinization temperatures of large numbers of varieties and selections in rice breeding programs.

Batcher *et al.* (1956, 1957) classified rice varieties using cooked samples of whole grain milled rice. Color, cohesiveness, flavor, degree of doneness, and amount of water absorbed were recorded. Rexoro, Texas Patna, Bluebonnet 50, and Century Patna 231 all absorbed more water during cooking and were less sticky than Caloro and Colusa. They concluded that grain type appeared to be associated with water absorption but noted some overlapping. Residual cooking liquids from most of the long-grain varieties appeared to have less total solids and starch than the short- and medium-grain varieties. However, Century Patna 231 and Toro had higher amounts of total solids in residual cooking liquids than did other long-grain varieties.

Williams *et al.* (1958) made comparisons between the amylose contents of different varieties using a modification of the colorimetric method of McCready and Hassid (1943). Amylose content was found to vary from 12.9% for Century Patna 231, to 23.4% for Texas Patna, both long-grain varieties. The short- and medium-grain varieties Zenith, Magnolia, Nato, and Caloro all had below

15% amylose.   Toro, a long-grain variety, also showed a low amylose content.

Soaking milled rice grains in dilute solutions of potassium hydroxide has been used by several investigators as a means of detecting varietal differences in milled rice.   Warth and Darafsett (1914) soaked short- and medium-grain Asiatic varieties in different concentrations of potassium hydroxide.   They found 2 main groups based on resistance of the rice grains to the alkali solution and, in the case of 3 varieties, associated these differences with gelatinization temperatures.   Jones (1938) divided US rice varieties into three groups and concluded that cooking behavior of varieties disintegrating into intermediate and clear masses was better than that of varieties showing opaque masses.

Little *et al.* (1958) classified 25 rice varieties from each of 4 geographical locations in the United States by soaking milled grains in a 1.7% potassium hydroxide solution.   Differences were based on the degree of spreading and clearing of grains, using a 7-point numerical scale.   They found that Century Patna 231, a long-grain variety, and Early Prolific, a medium-grain variety, showed the least amount of spreading and clearing and were usually rated 1 and 2 on the numerical scale.   The long-grain varieties were found to give variable reactions.   The short- and medium-grain varieties, except for Early Prolific, were more uniform in reaction, showed more spreading and clearing, and were usually scored 6 and 7 for spreading and 5 to 7 for clearing.

A cooking and soaking test reported by Halick and Keneaster (1956) gave marked varietal differences in the degree of longitudinal splitting of the grains.   The rice was cooked for 20 min and soaked overnight in petri dishes.   It was concluded that varieties exhibiting the greatest degree of longitudinal splitting showed the greatest degree of stickiness when cooked.

Amylose content can be estimated by the iodine-blue staining method, a dark-blue coloration being regarded by some investigators as necessary for good cooking performance.   The temperature of gelatinization is determined by observing granule swelling and loss of birefringence, or by cooking in the Brabender Amylograph.   Other methods useful for determining gelatinization characteristics are the alkali method, uptake of water at 170° and 180°F, and the heat alteration method.   The pasting behavior is measured with the Amylograph.

Some other tests which can be used to evaluate the cooking quality of rice are the soaking test, in which cooked rice is soaked overnight in cold water and the extent of kernel splitting judged, and the parboiling test for determining the suitability for canning and similar applications (Adair 1961).

Varieties which are (1) high in amylose, (2) gelatinize at intermediate temperatures, and (3) have a lower viscosity at 201°F than when cooled to 122°F are suitable for canning, quick cooking processing, and the like. The kernels of such varieties, usually long-grain types, show little splitting of the kernels in the soaking and parboiling tests and they are dry and fluffy when cooked.

Medium- and short-grain rice varieties grown in the United States are lower in amylose and have a lower gelatinization temperature than the long-grain varieties. They usually cook into a rather glutinous mass. They are acceptable to certain groups as table rices, and are satisfactory as carbohydrate sources in brewing, and for flaking and puffing (as for dry breakfast cereals). They generally gelatinize in the 142° to 150°F range.

When intended for use in baby food formulations, special precautions are taken to insure that the rice is free from filth, weed seeds, insect fragments, and the like (Kelly 1961). The gelatinization temperature affects the yield and the quality of the product in the manufacture of the flaked drum-dried rice cereal sold as baby food. The lower the gelatinization temperature, the better the final product. Short- and medium-grain rices are preferred.

For cereals such as oven-puffed rice, the manufacturers demand uniformity in variety, grain size, and moisture content. The kernels should be fully-milled, translucent, and have a fat content of less than 0.5% (Littlejohn 1966). Whole, plump kernels of the medium-grain varieties are preferred. Translucent kernels are thought to indicate uniform cohesiveness when cooked and to provide a better surface texture for intermediate handling, while chalky rice produces stickiness due to the penetration of the soft starch by the flavoring agents with resultant cracking and breaking of the kernels. Proper aging of the rice tends to result in a uniform response to processing, as contrasted to fresh rice which is susceptible to checking and may vary from lot to lot in its response. The specification for low fat content is a precaution against rancidity development in the packaged cereal.

Somewhat different qualities are desired for rice to be used for gun-puffed cereal. Either milled or parboiled rice can be used although they give different types of products. Milled rice should be very clean, and free from chalky kernels, broken rice, and any kind of coating. The major specifications for parboiled rice to be used in

gun-puffing include limitations on moisture content, number of ungelatinized and damaged kernels, and storage time. Contrasting
types are not desired. Long- and medium-grain rices are not suitable due to the shape and appearance of the puffed kernels. Protein
is not as much of a texture determinant as is the starch. There
appears to be an optimum amylose to amylopectin ratio for the best
puffing and expansion properties, but this optimum may vary from
crop to crop. There should be a minimum amount of germ remaining on the rice, since this turns black in the puffing processing
(Brockington 1966).

For the shredded type of cereal, one processer uses California No. 1
Second Head Milled Rice (Smith 1966). A very low content of
weeds, adobe balls, and other foreign materials is mandatory. In
general, a good quality, clean, broken rice of uniform grit size and
with a minimum of chalky grains, is specified. Hard flinty kernels
can be conveyed without the development of starch dust which leads
to an inferior product. Performance tests are needed to insure that
the rice will form a hard glazed surface and have consistent and pre-
dictable syneresis.

Grain for canned rice products must withstand extreme temperature abuse without disintegrating or becoming excessively mushy.
Only long-grain rice has been satisfactory, but different varieties of
these have shown great differences in canning stability. The best
varieties are Nira, Rexoro, and Texas Patna (Hagberg 1966). Rice
with low amylose content will always have poor stability, but, on the
other hand, not all rice with high amylose content will have the
necessary stability. Some lots within varieties fail to exhibit good
canning quality. Parboiling develops canning stability in some
rice. All of these uncertainities dictate the use of a good functional
test with conditions based on the times and temperatures encountered in the commercial process.

Particle size is of relatively little importance if the grain is to be
used as a brewing adjunct. Consequently, the finer particles from
the milling process usually find their way into this outlet. On the
other hand, uniformity of particle size is important. The gelatinization temperature must be low enough to allow the starch to be fully
cooked without destroying the malt amylase which has been added to
it. The viscosity should be low for ease in handling. A high fat
content adds to the viscosity and should be avoided; a specification
of less than 0.75% fat is typical (Hardwick 1966). Complete freedom from off-odors and off-flavors is essential, and sensory tests must
be relied upon to determine this characteristic, except for chemical

tests for herbicides and pesticides. For example, the methyl bromide concentration should be less than 10 ppm (Hays 1966).

A pilot brewing test is used by many quality control departments to evaluate the suitability of rice for brewing.

Flour made from short and medium-grain rice is preferred for use as a thickener in canned products, since the use of flour from long-grain rice leads to water separation, no doubt due to the retrogradation of the greater amount of amylose present in the long-grain varieties. Waxy rice provides flour having highly desirable properties for thickening frozen and canned sauces and gravies.

NUTRITIONAL ASPECTS

Although rice is one of the best cereals from a nutritional standpoint, it is not a completely adequate food. A large part of the world's population manages to survive and reproduce with rice furnishing about 80% of their total caloric intake, supplemented by relatively minor amounts of plant and vegetable sources of amino acids and vitamins together with minerals from various intentional or adventitious additives.

Rice protein is of fairly good quality, but it is present in inadequate amounts. As a result, rice alone cannot fulfill the protein requirements of man. If 1 gm of protein per kilogram of body weight per day is taken as the desirable intake for adults, some 50 gm would be required daily by a person of 110 lb weight. This is equivalent to between 600 and 700 gm of brown rice, or around 10 to 14 cups of cooked rice. Persons consuming rice *ad libitum* to the extent of 80% of their calories, do not ordinarily eat this much. A figure of somewhat less than ten cups has been given as a top level of voluntary consumption under stable conditions.

There are several ways of evaluating the adequacy of protein for human nutrition. Some of these are biological values based on feeding studies while chemical techniques can be used to compare the amino acid pattern with that of a standard such as whole egg protein. By any of these methods, rice protein is considerably inferior to egg, milk, and many other animal protein sources. The limiting amino acid is lysine, but isoleucine, threonine, tryptophan, and the sulfur-containing amino acids are also low.

Although the nutritional adequacy of rice lipids has not been extensively studied, it is known that linoleic acid is found in rice polish and rice oil, and they can be valuable sources of this essential dietary fatty acid. Rice seems to be low in allergenic factors.

Brown rice is deficient in fat-soluble vitamins and ascorbic acid.

Milled white rice is also very low in riboflavin and thiamine. Lack of thiamine leads to the deficiency disease beriberi. The thiamine content of rice can be improved by undermilling, parboiling, and enrichment. Adoption of one or more of these measures by suppliers to various parts of the world has caused a great reduction in the overall incidence of beriberi.

Enrichment

It has been known for a long time that hulled or brown rice and undermilled or hand-pounded rice are more nutritious (i.e., are less conducive to the development of deficiency diseases) than is milled or white rice. This is due to the concentration of vitamins and proteins in the embryo and the outer layers of the kernel. Most consumers greatly prefer milled rice, not only because of the more attractive appearance but also because of the blander flavor. When rice is a minor part of the diet, its nutritive defects have little effect on the well-being of the consumer. When rice is consumed at a rate sufficient to provide 80% of the caloric intake, as it is in many population groups, its adequacy from a nutritional standpoint is a critical point. The diets of these consumers can be improved either by educating them and their suppliers to change to brown rice, which, from the experience of other "educational" projects designed to change dietary habits, would probably be a long and costly procedure of doubtful success, or by improving the preferred form of their staple food with additives which do not substantially change its appearance, flavor, or texture. Enrichment and parboiling are attempts to follow the latter course.

In most cases, enrichment of raw rice has been restricted to the addition of vitamins and minerals. Supplementation with amino acids or proteins has been suggested but does not appear to have been put into practice anywhere on a commercial scale. Parboiling can be considered a form of enrichment in which some of the vitamins and minerals are transferred from the outer layers of the kernel (later removed by milling) to the interior, so that they are retained in the finished product.

Approximately 80% of the thiamine, 56% of the riboflavin, 65% of the niacin, 85% of the fat, 60% of the pantothenic acid, and 55% of the pyridoxine are removed from brown rice when it is milled. Most, but not all, of these constituents are retained if the rice is parboiled. Enrichment by additives, at least as it is practiced in the United States, aims only at raising the nutrients to their concentration in the original material and does not seek to raise them to a level higher than in the brown rice. Thiamine, niacin, iron, and (sometimes) riboflavin are the common enrichments. Enrichment by adding supplements is probably less costly than parboiling when the energy, labor, and loss expenses of the latter process are considered.

Some of the methods which have been used for adding dietary supplements to rice are (Hammes 1966): (1) Mixing with the kernels a powder containing the vitamins and minerals. It is necessary that the consumer be cautioned against washing the rice, for such an operation removes the enrichment. (2) Supplying a wafer which must be added to the rice at the time it is cooked. The problems which would be involved in securing the cooperation of distributors and consumers in this type of procedure are rather obvious, but such a method has been used successfully under military conditions. (3) Adding a small amount of rice grains which have been coated or impregnated with the enriching ingredients. The enriched grains are then protected against washing by covering them with a water insoluble coating. Rice carrier grains are commonly added at the rate of 1 lb to each 200 lb of milled rice. If riboflavin is included, the yellow color diffuses into the water when the coated kernels are boiled, causing unsightly blotches in the cooked rice.

BIBLIOGRAPHY

ACUNA, J. 1958. Information of general interest about rice. Administratión de Estabilización del Arroz. Bol. 5, 1–56.

ADAIR, C. R. 1934. Studies on blooming in rice. J. Am. Soc. Agron. 26, 965–973.

ADAIR, C. R. 1961. Food quality factors used in rice-breeding studies. Proc. Second Conf. Rice Utiliz., Albany, Calif. May 18–19, 1961.

ADAIR, C. R., and ENGLER, K. 1955. The irrigation and culture of rice. US Dept. Agr. Yearbook Agr. 1955, 389–394.

ANON. 1950. Storage of rough rice. Texas Agr. Expt. Sta. Prog. Rept. 1223.

ANON. 1951. A review of water seeding methods in Texas. Rice J. 54, No. 2, 11–12.

ANON. 1966. IR 8-288-3. High yielding IRRI selection. IRRI Reporter 2, No. 5, 1–2.

ANON. 1968A. Rice Millers Association rice acreage and production statistics. Rice J. 71, No. 7, 10–12.

ANON. 1968B. United States Standards for Rough Rice, Brown Rice, Milled Rice. US Dept. Agr. Consumer and Marketing Serv.

ATKINS, J. G. 1958. Rice diseases. US Dept. Agr. Farmers' Bull. 2120.

ATKINS, J. G., and ADAIR, C. R. 1957. Recent discovery of Hoja Blanca, a new rice disease in Florida, and varietal resistance tests in Cuba and Venezuela. Plant Disease Reptr. 41, 911–915.

ATKINS, J. G., BEACHELL, H. M., and CRANE, L. E.   1956.   Reaction of rice varieties to straighthead.   Texas Agr. Expt. Sta. Progr. Rept. *1865*.

BARR, H. T., and COONROD, L. G.   1951.   Present status of bulk drying and storage of rice on the farm.   Rice J. *54*, No. 8, 12–17.

BATCHER, O. M., DEARY, P. A., and DAWSON, E. H.   1957.   Cooking quality of 26 varieties of milled white rice.   Cereal Chem. *34*, 277–285.

BATCHER, O. M., HELMINTOLLER, K. F., and DAWSON, E. H.   1956. Development and application of methods for evaluating cooking and eating quality of rice.   Rice J. *59*, No. 12, 4–9.

BEACHELL, H. M.   1943.   Effect of photoperiod on rice varieties grown in the field.   J. Agr. Res. *66*, 325–340.

BEACHELL, H. M., ADAIR, C. R., JODON, N. E., DAVIS, L. L., and JONES, J. W.   1938.   Extent of natural crossing in rice.   J. Am. Soc. Agron. *30*, 743–753.

BEACHELL, H. M., and HALICK, J. V.   1957.   Breeding for improved milling, processing and cooking characteristics of rice.   Intern. Rice Commun. News Letter *6*, No. 2, 1–7.

BOWLING, C. C.   1956.   Control of the stink bug and grasshoppers on rice.   Texas Agr. Expt. Sta. Progr. Rept. *1900*.

BOWLING, C. C.   1957.   Seed treatment for control of the rice water weevil.   J. Econ. Entomol. *50*, 450–452.

BOWLING, C. C.   1968.   Rice water weevil resistance to Aldrin in Texas. J. Econ. Entomol. *60*, 1027–1030.

BROCKINGTON, S. F.   1966.   Puffed rice products.   Proc. Natl. Rice Utiliz. Conf., New Orleans, La. Apr. 5–6, 1966.

CHANG, T. T.   1968.   Personal communication.   IRRI, Manila, Philippines.

CHATTERJEE, D.   1948.   A modified key and enumeration of the species of *Oryza* Linn.   Ind. J. Agr. Sci. *28*, 185–192.

CHEANEY, R. L.   1955.   Effect of time of draining of rice on the prevention of straighthead.   Texas Agr. Expt. Sta. Progr. Rept. *1744*.

CRALLEY, E. M.   1949.   White-tip of rice.   Phytopathology *39*, 5.

DAVIS, L. L.   1950.   California rice production.   Calif. Agr. Exten. Serv. Circ. *163*.

DAVIS, L. L., and JONES, J. W.   1940.   Fertilizer experiments with rice in California.   U S Dept. Agr. Tech. Bull. *718*.

DAVIS, W. C.   1954.   Responses of rice to some herbicides.   Texas Agr. Expt. Sta. Progr. Rept. *1678*.

DAVIS, W. C.   1955.   Weed killers and rice.   Texas Agr. Expt. Sta. Progr. Rept. *1812*.

DEATHERAGE, W. L., MACMASTERS, M. M., and RIST, C. E.   1955. A partial survey of amylose content in starch from domestic and foreign varieties of corn, wheat, and sorghum, and from other starch-bearing plants.   Trans. Am. Assoc. Cereal Chemists *13*, 31–42.

DOUGLAS, W. A., and INGRAM, J. W.   1942.   Rice field insects.   US Dept. Agr. Circ. *632*.

EFFERSON, J. N.   1952.   The Production and Marketing of Rice.   Rice J., New Orleans, La.

EVATT, N. S., and WEIHING, R. M.   1957.   Fertilizer requirements for rice in rice-pasture rotations.   Texas Agr. Expt. Sta. Progr. Rept. *1948.*

FINFROCK, D. C., and MILLER, M. D.   1958.   Establishing a rice stand. Calif. Agr. Exten. Serv. Leaflet *99.*

GHOSE, R. L. M., GHATGE, M. B., and SUBRAHMANYAN, V.   1956.   Rice in India.   Indian Council Agr. Res., New Delhi, India.

GREEN, B. L., and WHITE, J. H.   1963.   Comparison of three selected rice rotations in Eastern Arkansas.   Arkansas Agr. Expt. Sta. Bull. *664.*

HAGBERG, E. C.   1966.   Canned rice products.   Proc. Natl. Rice Utiliz. Conf., New Orleans, La.   Apr. 5–6, 1966.

HALICK, J. V., and KELLY, V. J.   1959.   Gelatinization and pasting characteristics of rice varieties as related to cooking behavior.   Cereal Chem. *36,* 91–97.

HALICK, J. V., and KENEASTER, K. K.   1956.   The use of a starch-iodine-blue test as a quality indicator of white milled rice.   Cereal Chem. *33,* 315–319.

HAMMES, P. A.   1966.   Enrichment of rice.   Proc. Natl. Rice Utiliz. Conf., New Orleans, La.   Apr. 5–6, 1966.

HARDWICK, W. A.   1966.   Anheuser-Busch's use of rice as a brewing adjunct.   Proc. Natl. Rice Utiliz. Conf., New Orleans, La. Apr. 5–6, 1966.

HAYS, W. E.   1966.   Characteristics desired in rice for Adolph Coors Company.   Proc. Natl. Rice Utiliz. Conf., New Orleans, La.   Apr. 5–6, 1966.

HENDERSON, S. M.   1966.   Deep-bed grain drying on the ranch with unheated air.   Calif. Agr. Expt. Sta. Exten. Serv. Leaflet *103.*

HILDRETH, R. J.   1955.   An economic evaluation of on-farm drying and storage of rice in Texas.   Texas Agr. Expt. Sta. Progr. Rept. *1821.*

HODGES, R. J., Jr.   1957.   Rice—a big business on the gulf coast prairie.   Texas Agr. Exten. Serv. Bull. *B-782.*

ISELY, D., and SCHWARDT, H. H.   1934.   The rice water weevil.   Arkansas Agr. Expt. Sta. Bull. *299.*

JONES, J. W.   1936.   Improvement in rice.   US Dept. Agr. Yearbook Agr. *1936,* 415–454.

JONES, J. W.   1938.   The "alkali test" as a quality indicator of milled rice.   Am. Soc. Agron. *30,* 960–967.

JONES, J. W.   1943.   Upland rice.   US Dept. Agr. Multigraphed Circ.

JONES, J. W., ADAIR, C. R., BEACHELL, H. M., JODON, N. E., and WILLIAMS, A. H.   1953.   Rice varieties and their yields in the United States 1939–50.   US Dept. Agr. Circ. *915.*

JONES, J. W., DAVIS, L. L., and WILLIAMS, A. H.   1950.   Rice culture in California.   US Dept. Agr. Farmers' Bull. *2022.*

JONES, J. W., DOCKINS, J. O., WALKER, R. K., and DAVIS, W. C.   1952. Rice production in the southern states.   US Dept. Agr. Farmers' Bull. *2043.*

JONES, J. W., and LONGLEY, A. E.   1941.   Sterility and aberrant chromosome number in Caloro and other varieties of rice.   J. Agr. Res. 62, 381–399.

KELLY, V.   1961.   Properties of rice products desirable for baby food formulations.   Proc. Second. Conf. Rice Utiliz., Albany, Calif.   May 18–19, 1961.

KESTER, E. B.   1959.   Rice processing.   In The Chemistry and Technology of Cereals as Food and Feed.   S. A. MATZ (Editor).   Avi Publishing Co., Westport, Conn.

KRAMER, H. A.   1948.   Drying combined rice.   Bur. Plant Ind., Soils and Agr. Eng.   US Dept. Agr. Rept.

LAUDE, H. H., and STANSEL, R. H.   1927.   Time and rate of blooming in rice.   J. Am. Soc. Agron. 19, 781–787.

LINDEMANN, E.   1951.   Rice starch.   Stärke 3, 141–150.

LITTLE, R. R., HILDER, G. B., and DAWSON, E. H.   1958.   Differential effect of dilute alkali on 25 varieties of milled white rice.   Cereal Chem. 35, 111–126.

LITTLEJOHN, J. P.   1966.   Production of and characteristics desired in rice for Rice Krispies and Special K.   Proc. Natl. Rice Utiliz. Conf., New Orleans, La.   Apr. 5–6, 1966.

LUGAY, J. C., and JULIANO, B. O.   1964.   Fatty acid composition of rice lipids by gas-liquid chromatography.   J. Am. Oil Chemists' Soc. 41, 273–275.

McCREADY, R. M., and HASSID, W. Z.   1943.   The separation and quantitative estimation of amylose and amylopectin in potato starch.   J. Am. Chem. Soc. 65, 1154–1157.

MONCRIEF, J. B., and WEIHING, R. M.   1950.   Rapid, low-cost conversion from rice to improved pastures.   US Dept. Agr. Bull. 729.

MORRISON, S. R., DAVIS, W. C., and SORENSON, J. W., JR.   1954.   Bin drying of rice at the Rice-Pasture Experiment Station, 1953–54.   Texas Agr. Expt. Sta. Progr. Rept. 1670.

MULLINS, T., and SLUSHER, M. W.   1951.   Comparison of farming systems for large rice farms in Arkansas.   Arkansas Agr. Expt. Sta. Bull. 509.

PALMIANO, E. P., ALMAZAN, A. M., and JULIANO, B. O.   1968.   Physicochemical properties of protein of developing and mature rice grain.   Cereal Chem. 45, 1–12.

RAMIAH, K., and GHOSE, R. L. M.   1951.   Origin and distribution of cultivated plants of South Asia—Rice.   Indian J. of Genetics and Plant Breeding 11, No. 1, 7–13.

RAMIAH, K., and RAO, M. B. V.   1953.   Rice Breeding and Genetics.   Indian Council of Agr. Res., Scientific Monograph 19.

REYNOLDS, E. B.   1954.   Research on rice production in Texas.   Texas Agr. Expt. Sta. Bull. 775.

SAMPATH, S., and RAO, M. B. V.   1951.   Interrelationships between species in the genus Oryza.   Indian J. of Genetics and Plant Breeding 11, No. 1, 14–17.

SCHOCH, T. J.   1967.   Properties and uses of rice starch.   In Starch: Chemistry and Technology, Vol. II.   R. L. WHISTLER, and E. F. PASCHALL (Editors).   Academic Press, New York.

SMITH, R., JR.   1966.   Weeds and their control in rice production.   US
Dept. Agr. Handbook *292.*

SMITH, R., JR.   1968.   Control of grass and other weeds in rice with
several herbicides.   Arkansas Agr. Expt. Sta. Rept. Series Bull. *167.*

SMITH, R. G.   1966.   Desirable rice characteristics for Ralston Purina
Company.   Proc. Natl. Rice Utiliz. Conf., New Orleans, La.   April
5–6, 1966.

SORENSON, J. W., JR., and DAVIS, W. C.   1955.   Drying and storing
rough rice in farm storage bins, 1954–55.   Texas Agr. Expt. Sta. Progr.
Rept. *1819.*

SREENIVASAN, R.   1968.   Component fatty acids and composition of
some oils and fats.   J. Am. Oil Chemists' Soc. *45,* 259–265.

TAYLOR, T. C., and NELSON, J. M.   1920.   Fat associated with starch.
J. Am. Chem. Soc. *42,* 1726–1738.

TISDALE, W. B., and JENKINS, J. M.   1921.   Straighthead of rice and
its control.   US Dept. Agr. Bull. *1212.*

WALKER, R. K., and MIEARS, R. J.   1957.   The coastal prairies.   US
Dept. Agr. Yearbook Agr. *1957,* 531–534.

WARTH, F. J., and DARAFSETT, D. B.   1914.   Disintegration of rice
grains by means of alkali.   Agr. Inst. Pusa (India) Bull. *38.*

WASSERMAN, T., MILLER, M. D., and GOLDEN, W. G., JR.   1966.   Heated
air drying of California rice in column dryers.   Calif. Agr. Expt. Sta.
Exten. Serv. Leaflet *184.*

WEIHING, R. M., MONCRIEF, J. B., and DAVIS, W. C.   1950.   Yearlong
grazing in the rice-pasture system of farming.   Texas Agr. Expt. Sta.
Progr. Rept. *1280.*

WHITEHEAD, F. E.   1954.   Test on insecticidal control of rice water
weevil.   J. Econ. Entomol. *47,* 677–680.

WILLIAMS, R. E.   1955.   Weeds in rice.   Rice J. *58,* No. 13, 18–19;
59, No. 1, 8–9; No. 2, 8; No. 3, 8–9; No. 4, 14–15.

WILLIAMS, V. R.   1961.   Rice Proteins.   I. Proc. Second Conf. Rice
Utiliz. Albany, Calif.   May 18–19, 1961.

WILLIAMS, V. R., WU, W. T., TSAI, H. Y., and BATES, H. G.   1958.
Varietal differences in amylose content of rice starch.   J. Agr. Food
Chem. *6,* 47.

WYCHE, R. H.   1941.   Fertilizer for rice in Texas.   Texas Agr. Expt.
Sta. Bull. *602.*

CHAPTER 8

S. A. Matz

Millet, Wild Rice,
Adlay, and Rice Grass

INTRODUCTION

The preceding chapters in this book are concerned with the seven grains which are most important to the economy of the United States. In the present chapter, four cereal grains of minor significance will be briefly discussed. Two of these are plants which are currently grown and the products sold for food or feed, though in small quantities. One of the others is primarily of historical interest, while the fourth has never been used in this country though it has interesting possibilities.

It might be worthwhile to remind the reader of the predetermined limitations of this volume. Only cereal grains are considered, and so buckwheat (an herbaceous plant), soybeans (a legume), and other nongrasses bearing seeds which are commonly processed and consumed similarly to cereal grains are not included. Furthermore, the forage aspects of cereal plants are touched upon only in passing. The greatest emphasis has been placed upon practices followed in the United States, although there has been no intention of ignoring the more important variations of these procedures found in other countries.

MILLET

The name of millet has been applied, at different times and at different places, to a wide variety of cereal grains. *Sorghum vulgare* has been erroneously called pearl millet and in some parts of the world all varieties of sorghum and millet are designated by the latter name. However, the grains most generally recognized as millets belong to two tribes of the grass family, the Chlorideae and the Paniceae, while sorghums belong to one genus (*Sorghum*) in one tribe (Andropogoneae) of the grass family.

The tribe Chlorideae includes *Eleusine coracana* as the only species of economic importance (Hitchcock 1950). This plant, called variously African, ragi, or finger millet, is grown very extensively in India where the grain is used as human food. The tribe Paniceae includes several species grown for food and feed in various parts of the world. The most important of these species will be described below.

224

Setaria (Chaetochloa) italica includes many varieties such as foxtail millet, German millet, Hungarian grass, Siberian millet, and Kursk millet which are grown in China or Russia for human use, but in the United States only for forage. *Pennisetum typhoideum*, pearl

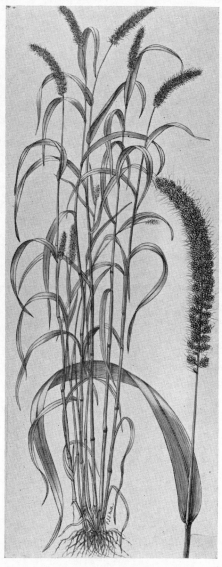

Courtesy of US Dept. Agr.

FIG. 45. FOXTAIL MILLET (*Setaria italica*)

millet, is extensively cultivated in Egypt and in tropical Asia—particularly India—as a food cereal. *Echinochloa decompositum* is the Australian millet, grains of which are said to be used as food by the aborigines of that continent. Other species of *Echinochloa* are grown for food in tropical Africa and in South America. The only genus of any economic importance in the United States is *Panicum*, and *Panicum miliaceum*, proso or common millet (also called broomcorn millet and hog millet) makes up the largest crop.

The origin of *Panicum miliaceum* is unknown, but it is probably a native of Egypt and Arabia. It has been cultivated in Asia Minor and southern Europe since prehistoric times. Millet kernels were found among the habitations of the Swiss Lake-Dwellers. It is a small, erect annual, reaching a height of 3 to 4 ft and possessing bristly, much branched panicles. The grain is about 3 mm long by 2 mm broad, and is usually inclosed in the shining, hard, flowering glume and palea. The glumes may be red or yellow, or any shade of gray. Three botanical varieties are recognized and designated as *effusum*, *contractum*, and *compactum*.

Soil preparation and cultivation procedures are little different from those used for other cereals. The seed bed is prepared by plowing, harrowing, and cultipacking or otherwise firming the seedbed. The seed can be sown either by drilling or by broadcasting, though drilling is used most extensively. It is sown during the period from late May to early July.

About 60 to 75 days after seeding, millet is ready for harvesting. Ripening is not uniform, and frequently it is found that the grain in earlier panicles or in the tops of the heads are ripe and have shattered before the lower seeds and later panicles are completely mature. Because of the irregular ripening, much seed is lost by direct combining and the usual practice is to cut the crop by a swather. The windrows formed in this manner are allowed to cure, and then combined. The crop can also be harvested with a grain binder and placed in shocks to cure. Thereafter, it is handled like wheat or other small grains. Up to 20 bu of seed per acre can be obtained under favorable conditions. The weight of the seed is from 48 to 60 lb per bushel.

Although millets constitute one of the world's most important groups of food plants, their cultivation is restricted mostly to the Eastern Hemisphere, and, in particular, to regions of rather primitive agricultural practices and high population density. During the Medieval era, it was one of the principal grains produced in Europe. The reasons for its decline in popularity are rather obvious.

Meal and other preparations of millet have a strong taste which is not generally preferred by persons having access to blander grains. Furthermore, the meal cannot be made into leavened bread, so it must be consumed as a gruel or made into flatbread.   Higher yielding grasses are available for feed purposes, except where certain rather unusual combinations of soil and climate exist.   For example, oats generally outyield millet, except for some areas where sandy soil, hot growing weather, and scanty rainfall combine to make the environment more favorable for millet.

In comparison with wheat, corn, and oats, the production of millet in the United States seems ridiculously small.   Most of it is grown in North Dakota, South Dakota, and Colorado, and is of the proso variety.   The seed is used in the United States principally for poultry feed, but some is used for stock feed, and a small amount is used in commercial feed mixtures for pet birds such as canaries and parakeets.   Occasionally, whole panicles of the plant are sold in pet food stores for hanging in bird cages.

The plant can be grown as far north as 54° N latitude.   It does not grow well anywhere until the soil is warm.   In northern states, seed may not be sown until late June or early July.   The seed deteriorates rapidly, and it is generally considered undesirable for use after more than two years of storage, because of the low rate of germination.

As a general rule, fertilizers are not used with millet in the Great Plains states.   Farther east, nitrogen and phosphate applications have caused increased yields, but it appears to be more economical to use the fertilizer on other crops in the rotation.

Twenty-five to 40 lb of seed per acre are recommended for areas of ample rainfall, while 15 to 20 lb per acre should be used in drier localities.   The smaller amounts of these ranges are more appropriate where seed production is intended and the larger amounts are suitable for forage planting.

In spite of its unpopularity as food, millet has a nutritive value comparable to that of other cereals.   The "average composition" of several grains including millet is shown in Table 54.   It can be seen that both proso and foxtail millet are somewhat higher in protein than rice, sorghum, corn, and oats.   The biological value of the protein for humans has apparently not been determined.   Millet has a high percentage of indigestible fiber because the seeds are inclosed in hulls which are not removed by ordinary processing methods.   It is a good source of thiamine, and probably contains appreciable amounts of the other B vitamins.   Similar varieties of

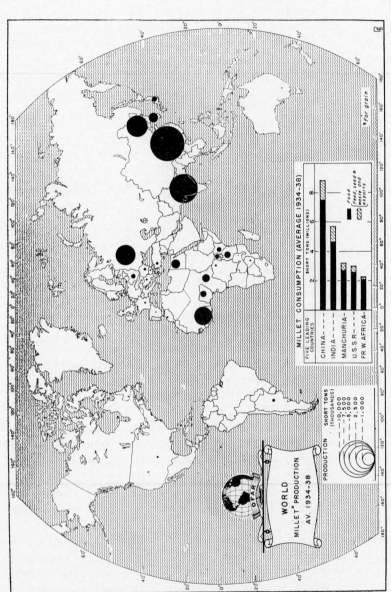

Courtesy of US Dept. Agr.

Fig. 46. Distribution of the World's Millet Production

TABLE 54

AVERAGE COMPOSITION OF MILLET AND OTHER GRAINS[1]

Grain	Average Analysis (%)				
	Protein	Fat	Fiber	Carbohydrate, Except Fiber	Ash
Proso millet	11.7	3.3	8.1	64.2	3.4
Foxtail millet	12.1	4.1	8.6	60.7	3.6
Sorghum	11.3	2.9	2.2	71.3	1.7
Adlay[2]	13.6	6.1	8.4	58.5	2.6
Rice (rough)	7.9	1.8	9.0	64.9	5.2
Wild rice[3]	14.1	0.7	1.5	74.4	1.2
Oats	12.0	4.7	10.6	60.2	3.6
Corn	9.7	4.0	2.3	71.1	1.4
Wheat	13.2	1.9	2.6	69.9	1.9
Barley	11.8	2.0	5.7	68.0	2.9

[1] Many of these data were obtained from Morrison (1948).
[2] Leme de Rocha (1950).
[3] Nelson and Palmer (1942).

millet grown in different parts of the world do not seem to vary greatly in composition (Anon. 1937; Wu 1928; Morrison 1948).

WILD RICE

Wild rice (*Zizania aquatica*) is botanically rather far removed from true rice, though both are classified in the tribe Oryzeae. In appearance the plant does not greatly resemble rice, and the grain itself has few similarities with its cultivated namesake. Evidently the aquatic habit of wild rice led to its name. The plant is also called Tuscarora rice or Indian rice.

Zizania aquatica is an annual grass, usually 5 to 10 ft tall, which bears 4 to 6 leaves 1 to 4 cm in width and up to 65 cm long. The panicles are usually 30 to 50 cm long and the branches 15 to 20 cm long. The seeds are covered with palea and lemma when they fall from the panicle. Upon removing the palea and lemma, the black kernel is revealed to have the shape of a cylinder with rounded ends. The kernel is about 10 to 20 mm long and 0.5 to 1.0 mm in diameter (Anon. 1948).

Wild rice is found on mud flats over almost all the eastern half of the United States and in southern Canada eastward of Lake Winnipeg. Related species are cultivated in various other portions of the world. Minnesota is the largest producer of wild rice, and Wisconsin is second. The harvest in Canada is usually considerably less than in the United States. Probably the total value of all wild rice entering commercial channels in the Western hemisphere amounts to less than two million dollars annually.

It is said that seeds of wild rice are sometimes planted in marshes and on grain preserves, but the stands which are of the greatest economic importance have become established spontaneously. Except in cases where heavy attacks by parasites occur, ample reseeding occurs to keep the wild beds established year after year, even when the bulk of the grain is harvested. The grains, inclosed by the palea and lemma, become detached from the parent inflorescence in late autumn and fall into the water. Eventually the grain sinks into the muddy bottom and remains in a dormant condition throughout the winter (Brown and Scofield 1903).

The time of germination depends upon the water temperature and may vary in different localities from the latter part of March to the beginning of June. About a month after germination, the stalk rises above the water. Shortly thereafter, the panicles appear and the plants begin to flower. The stalks continue to grow until the grains are almost mature.

As a rule, wild rice must be harvested during the first two weeks of September or else most of the grain will fall into the water and be lost. The ripening process is not simultaneous for all grains in a panicle and the mature kernels are easily dislodged and lost in the water, so several harvests are necessary if the greatest yield is to be obtained from a rice bed.

Wild rice is harvested by Indians, mostly by members of the Chippewa tribe. Canoes and flat-bottomed boats, manned usually by two people, are used for gathering the grain (Stevens 1952). The boats are poled through the rice beds by one of the occupants while the other draws the ripe heads of grain over the side of the boat with a stick. When the panicle is in position, it is struck sharply with another stick, dislodging the grain into the boat. Perhaps a bushel (30–60 lbs) of wild rice can be gathered in an hour by this method. Large scale mechanical harvesters have also been put into use in 1 or 2 areas, but the unfavorable condition of the terrain over which they must operate greatly hampers their use. Furthermore, mechanical harvesting has been banned by law in Minnesota since 1939.

After harvesting, the grain is dried for 2 to 3 days in the open air and then it is parched and the hulls removed by pounding. Data on the changes in moisture content during drying and parching do not seem to be available although the marketed product frequently contains 7–10% moisture. Parching is often done with crude equipment, frequently a makeshift container supported over a wood fire. Large processers pass the grain through rotating cylinders (engine-

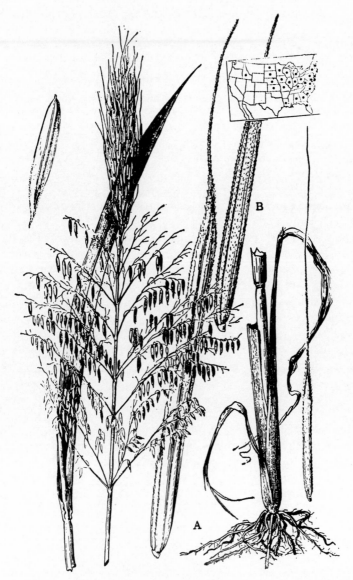

Courtesy of US Dept. Agr.

Fig. 47.   Zizania Aquatica var. Interior

A. Plant, ×¹/₂; Pistillate Spikelet, ×2; Second View, ×5. B. Pistillate Spikelet, ×5.

driven) situated over heating units. Control of the processing times and temperatures is a skill dependent upon an accurate judging of the texture and color of the product. Excessive heat causes popping of the rice.

After parching, the hulls of the wild rice are removed by subjecting the grain to impact. Formerly, the Indians did this by stamping on the grain, but mechanical devices are used now. A common type of apparatus consists of cylinders with rotating cores of rubber-covered spokes. Suction devices draw off the hulls as they are loosened. The kernels then pass over screens which remove broken grain and the pieces of hulls.

The wild rice of commerce is strictly a gourmet food. It is by far the most expensive cereal on the market, usually selling for over $3.00 per pound. In American cuisine, wild rice is used principally in combination with poultry and game birds, either as a stuffing or as a side dish. In such a context, discussion of the nutritive value of a food is rather academic since it is hardly likely that it will ever make up a significant part of anyone's diet. However, a comparison of the composition of wild rice with that of other cereals is included in Table 54 for the sake of completeness. Vitamin content of wild rice is of the same order as that of whole wheat in most respects, except that the former cereal contains considerably more vitamin B_2. Wild rice is a fair source of calcium and phosphorus (Nelson and Palmer 1942).

INDIAN RICE GRASS

Indian rice grass (*Oryzopsis hymenoides*) is a perennial which grows wild in many of the states of the western United States. Figure 48 gives an indication of the appearance of the plant. It has been described as a densely tufted bunchgrass growing from 1 to 2 ft tall. The leaves are slender and almost as long as the stems. The spreading panicle bears long pediceled spikeletes and lemmas with silky hairs.

Seed formation, when it occurs, results in the formation of seeds resembling millet. They are short and almost round in contour; they are dark, approaching black in color; and are covered profusely with white hairs. The seed bears a short awn.

The grain was frequently gathered by the Indians for use in much the same manner as they used wild rice (Anon. 1948). They were parched, or ground into meal and flour for making unleavened bread. Rice grass is now used almost exclusively for forage, and it is rather highly regarded for this purpose in areas where extensive stands of

Courtesy of US Dept. Agr.

Fig. 48.   Indian-Rice Grass

the native plant occur. Bohmont and Lang (1957) reported variations in the morphological characteristics and the palatability to
animals of some geographic strains of rice grass. It has been recommended for range reseeding (Verner 1956), but apparently is not
widely utilized in this connection at the present time. It competes
well on dry sandy soils and may be the predominant plant in sand
dune areas. As might be expected from this distribution, it is tolerant to drought. It is also relatively tolerant to high concentrations of minerals in the soil.

ADLAY

The available literature does not contain any references to the
utilization or growth of adlay or Job's tears (*Coix lacryma-Jobi*) in
the United States. The plant has been grown for its ornamental
properties in this country, but it is no longer very popular for this
purpose. It has been used for food in the Orient for thousands of
years and a variety selected for high protein content was introduced
into Brazil some years ago. Several plantations now exist in the
Sao Paulo area of Brazil (von Schaafhausen 1952).

The plant is a robust branched grass. Most varieties are 4 to 6
ft tall, though some are shorter. Figure 49 gives an indication of
the appearance of the plant. The inflorescences are made up in
part of hard, hollow, beadlike structures which are globular or somewhat pear-shaped. The inflorescence, one of which develops at the
end of each peduncle from a leaf sheath, exhibits a wide range of colors, from white through yellow, red, and purple to brown. *Coix* is
monoecious, and the staminate spikes project from an orifice on the
tips of certain of the beadlike structures, but the pistillate flowers
are inclosed with only the styles projecting.

Wester (1921) studied the nutritional aspects of the many varieties of adlay grown in the Philippines. He concluded that the protein content was similar to that of wheat, but that the biological
value of adlay protein was higher. The composition of adlay grain
is shown in Table 54. Several methods of preparation of the grain
were tried by Wester. He found that it could be hulled by machin-
ery designed for hulling rice. The grain could also be ground and
made into excellent biscuits (crackers). Von Schaafhausen (1952)
made bread and biscuits out of a mixture of 30% adlay flour and
70% wheat flour. The Chinese use the grain in soups, and in Japan
it is made into a fermented beverage (Kogama and Yamato 1955).
Experiments indicated that adlay meal can be substituted for wheat
bran and middlings in a balanced ration for chickens. Hog feeding
tests also showed good results.

Courtesy of US Dept. Agr.

FIG. 49.   ADLAY OR JOB'S TEARS

Apparently one of the chief reasons why this cereal has not achieved greater popularity is that it requires a long growing season. An improved variety promoted by von Schaafhausen overcomes this difficulty and has the additional advantage of high yield.   It ripens in 5 months or less, as opposed to about 6 or 7 months for the other varieties.   The yield is said to be greater than that of rice and in many instances greater than that of corn.   It has been grown in temperate regions of Brazil as well as in tropical and subtropical countries.

In growing adlay, the soil is prepared as for other cereals. In Brazil, at least, the proper time for sowing is about the same as for corn. Sowing is done by machine in rows which are about two feet to a yard apart, depending upon the fertility of the soil. The plants germinate slowly and it is essential to keep the weeds under control during the first month. Harvesting has been done by hand in most of the plantations, but it is said that the seed can be mechanically harvested by methods used for barley.

BIBLIOGRAPHY

ANON. 1937. The food and nutrition of African natives. International Institute of African Languages and Cultures. Memo. *13*.

ANON. 1948. Grass. US Dept. Agr., Yearbook Agr.

ANON. 1956. Nutritional Data, 3rd Edition. H. J. Heinz and Co., Pittsburgh, Penna.

BOHMONT, B. L., and LANG, R. 1957. Some variations in morphological characteristics and palatability among geographic strains in Indian rice grass (*Oryzopsis hymenoides*). J. Range Management *10*, 127–131.

BROWN, E., and SCOFIELD, C. S. 1903. Wild rice, its uses and propagation. US Dept. Agr., Bur. Plant Ind., Bull. *50*.

HITCHCOCK, A. S. 1950. Manual of Grasses of the United States. US Dept. Agr. Misc. Publ. *200*.

KOGAMA, T., and YAMATO, M. 1955. Studies on the constituents of coix species. J. Pharm. Soc. Japan *75*, 699–704.

LEME DE ROCHA, G. 1950. Analyses of adlay. Colheitas e Mercados *6*, No. 1, 12–13.

MORRISON, F. H. 1948. Feeds and Feeding. Morrison Publishing Co., Ithaca, N.Y.

NELSON, J. W., and PALMER, L. S. 1942. The thiamine, riboflavin, nicotinic acid, and pantothenic acid constituents of wild rice (*Zizania aquatica*). Cereal Chem. *19*, 539–546.

STEVENS, T. A. 1952. Wild rice-Indian food and a modern delicacy. Econ. Botany *6*, 107–142.

VERNER, J. E. 1956. Value of Indian rice grass (*Oryzopsis hymenoides*). J. Range Management *10*, 127–131.

VON SCHAAFHAUSEN, R. 1952. Adlay or Job's tears—A cereal of potentially greater importance. Econ. Botany *6*, 216–227.

WESTER, P. J. 1921. Nutritional aspects of adlay. Philippine Agr. Rev. *15*, 221–228.

WU, H. 1928. Nutritive value of Chinese foods. Chinese J. Phys., Rept. Series 1928, No. 1.

Index

LIBRARY
FLORISSANT VALLEY COMMUNITY COLLEGE
ST. LOUIS, MO

INVENTORY 1983

COMPLETED